THEY ARE ALREADY HERE

THEY ARE ALREADY HERE

UFO CULTURE AND WHY WE SEE SAUCERS

SARAH SCOLES

PEGASUS BOOKS

NEW YORK LONDON

THEY ARE ALREADY HERE

Pegasus Books Ltd.
148 W. 37th Street, 13th Floor
New York, NY 10018

Copyright © 2020 Sarah Scoles

First Pegasus Books edition March 2020

Interior design by Maria Fernandez

Library of Congress Cataloging-in-Publication Data is available.

ISBN: 978-1-64313-305-8

10 9 8 7 6 5 4 3 2 1

Printed in the United States of America
Distributed by W. W. Norton & Company

"I know the CIA would say
what you hear is all hearsay.
I wish someone would tell me what was right"
—Blink 182, "Aliens Exist"

CONTENTS

ONE

DA VINCI'S
GARAGE DOOR OPENER

I

've only seen a UFO once. And only for a second: It quickly turned into an IFO—an identified flying object. But the fleeting feeling that accompanied that fleeting unidentification stuck around for much longer.

It happened on August 21, 2017, about 20 miles outside of Jackson, Wyoming. I'd been camping for two nights already with friends. As such, we were dirty and tired and often too cold or too hot. A creek rushed behind our tents, and we used it to filter water and cool the beer we'd snagged from the nearby ski town's ostentatiously wood-beamed liquor store.

A hundred or so feet up the clearing, there was a guy in an RV who liked to shoot his gun at the mountainside. He didn't care that there was a total solar eclipse happening that day. In fact, he had driven up this rutted, rocky Forest Service road to get *away* from the phenomenon and from the swarm of wealthy tourists who'd invaded his town to see the moon cross in front of the sun in a pretty place.

SARAH SCOLES

We did care, though. Which was why we left before dawn that morning, heading up Cow Creek Trail toward a high point called Cream Puff Peak, 6 miles away and at nearly 10,000 feet of elevation. We wanted to be alone and closer to the edge of Earth's atmosphere so that we could feel—even though it was 7.5 billion times untrue—that we were the only people on the whole planet, the only ones who could see this celestial event. It would be *ours*.

We never found Cream Puff Peak, though, the trail seeming to twine differently from its path on the map. Instead, we settled onto an unnamed promontory a few minutes before Earth's only natural satellite started to slide in front of its only star. We ate cheese sticks and beef jerky and picked the chocolate out of trail mix, sticking our opaque eclipse glasses in front of our faces every so often to watch the sun's transmogrification into something *other*, as it dimmed and dimmed and dimmed, the moon biting Pac-Man chunks from it.

The disappearance took a while. And, to be honest, it was boring at first. But then the air changed. It seemed—although air does not have a color—to be yellower, like it was its own transition lens. It got colder. The colors of the pines, the subalpine grasses, and the sky itself seemed matte, although I hadn't thought of them as glossy before. The scene became perspectiveless, depthless, like a flat medieval painting where everything is right in front of you.

We each took out our jackets and shrugged them on. A few minutes later, we zipped them up. Our chatter—about how awesome jerky is, how Cream Puff was maybe that peak way across the valley (or maybe not), how far we'd come, how alone we were out here—had quieted. We sat down on the edge of the outcropping and looked up and around, silent except for an occasional "weird" or "wow," lost in our own internal experiences of the external universe.

Finally, the moment of totality—when the moon completely blocks the sun—was upon us. I wrapped the arms of my eclipse glasses around my ears and looked exactly where, normally, you're not supposed to. The last crescent of sunlight was disappearing, and

2

just before the moon notched and locked itself inside the sun's circle, a lens flare of light burst from the boundary. It's an event known as "the diamond ring."

"Whaaaaaaaaat?" I said out loud. I yelled at my companions to put their glasses on. "It's happening," I said.

The sky grew dark. A cold wind kicked up. The star that had risen and fallen every day of my life prior to that point wasn't there anymore. It had been replaced by something alien—which had, in turn, transformed the landscape into that of an exoplanet. Earth became a place I'd never lived but suddenly found myself, as if I'd been sucked up in a tractor beam and plunked down light-years away on someone else's home.

The sun's outer atmosphere, called the corona, which had existed *right there* the whole time but had remained invisible till now, wisped from the star's edges and licked at the center of the sky. Everything familiar felt deeply, deeply strange. During the few minutes of totality, its strangeness never became familiar.

I felt like I did on my first scuba dive, or that time in college when I did mushrooms and stood crying in front of a purple flowering bush because purple was such a beautiful and unlikely color: The world had always been this amazing and weird, right underneath regular reality. And I had only just realized it.

That's not my UFO story, although it probably primed me. The UFO showed up later that evening. When we descended from not-Cream-Puff, we drank our creek beers and sat around the dark fire pit that we weren't allowed to light up because Smokey said so. We relived the day ("Remember how you just said, 'Whaaaaaaaat?'") before moving on to a discussion of what we'd do in the event of not-quite-apocalyptic nuclear war, which at the time didn't seem over-the-top unlikely.

Finally, we stepped out from beneath our coniferous canopy to look up at the stars—all those other suns, orbited by all those other planets, whereon some other schmucks might be shivering and discussing the ineloquence of one of their companions during the eclipse of their own star.

It was dark out there: dark-dark. And it was clear, the high, dry air of western Wyoming putting few molecular walls between us and the view. Here, too, we stayed pretty quiet, letting the stars' light flow into our eyes while we each thought our own thoughts.

"There's a satellite," said my friend Tripp, a medical physicist who had recently declared that in the event of not-quite-apocalyptic nuclear war, he would go to the affected area and treat people. He pointed to a fast-moving dot tracing an arc between the stars.

Before any of us could respond, the dot grew brighter. Light started coming from its edge. Right as my brain registered that change, the brightness swept down. Then it became a beam, searching. In a second, it was pointed straight at us.

"Whaaaaaaaat?" we all said, involuntarily, in unison.

It knows we're here, was my first involuntary thought. Terrestrial explanations then came quick: It was a Forest Service helicopter. It was the Jackson Hole police's rotorcraft. It was the military, checking on the throngs. It was some rich asshole on some rich asshole tour.

But I knew—from the light's prior movement among the stars—that it was not a helicopter. And for the second or so when I was in its spotlight, there was a part of me (a part of me I didn't really want to acknowledge) that believed it was *possible* that *maybe* this was it. Those extraterrestrial UFOs I'd long—with my skeptic-minded science brain—said didn't exist? They were *here* now, and they knew *I* was here, and they wanted me to know that they knew that I was here.

What if this, I thought, is the moment *right before* everything changes?

It's a way I always feel when something in the world seems off, not unlike the way it did during the eclipse. When no one is out early on a Sunday morning, a part of me entertains the possibility that, five minutes from then, I'll find out 90 percent of everyone died overnight of a fast-working flu or a rapture. When my car won't start right away on a frigid winter morning, I wonder if I'm about to realize that this is due to is a widespread EMP attack. When the *urgh-urgh-urgh* of an emergency broadcast comes on the radio, my lizard brain decides this is the time the announcer will say, "Yeah, it's aliens."

After all, most speculative or magical movies begin with a few *before* moments. Their characters live in the same grounded, rule-following world the audience does—until, all of a sudden, they don't. If you squint at things a certain way, it feels like that could happen to you too.

Anyway, I felt for a second, almost against my will, like maybe this was my magical *before*. The world has seemed more full of this feeling than usual recently, as we wonder if we stand on the precipice of some dystopian moment—the point at which divisions become irreparable, democracy ends, privacy totally evaporates, and world wars almost begin.

But just as quickly as we'd found ourselves standing in the seemingly sentient light, it was gone, the dot continuing on its path through the sky that now seemed stranger than ever, in this place that felt less *alone* than it had before.

What was that, what was that, what was that, we all said to each other. We ran through our terrestrial ideas and laughed at the extra-terrestrial one. We were both unnerved and enervated by the Not Knowing.

A minute later, I remembered what I knew: That Iridium communications satellites flare when they pass over, as the sun flashes against their shiny surfaces. If you're lucky enough to be under their path at the right angle, it *always* looks like they're staring right at you. The sight was a coincidence of timing and geography, just like the eclipse.

Good, I thought, satisfied. An answer. I wouldn't have to tell anyone I'd seen something I couldn't explain. I wouldn't have to live with uncertainty. I wouldn't have to believe anything about anything. And this is the way I like it: Belief, and the idea that something might lie beyond investigation, feel anathema to me. Still, I had to admit that a current ran through my circulatory system when I thought that maybe—just maybe—I was wrong. After all, wasn't it beautiful, briefly, to think that underneath the world I knew well was another that I didn't get at all?

That seed had been planted, but it didn't turn into a purple flower, merely into a smile-inducing, head-shaking memory. Then, on December 16, 2017, my phone buzzed against my thigh. I was sitting in the passenger seat of a car, bundled up against the Colorado cold. My friend Tasha, an editor who'd also been around for the Iridium flare, and I were on the way to cross-country ski up an isolated valley a couple hours from Denver, where I live. "The Pentagon has acknowledged a secret program to investigate UFOs," read the *New York Times* breaking news alert.

I put my phone back inside my snow pants and thought, "Duh. Big deal." *Of course* the Pentagon had a UFO investigation program. UFOs are just what the words behind their acronym mean: unidentified flying objects. The connotation of the term in common parlance is, of course, extraterrestrial, but its denotation is not. It simply means something in the sky that the seer cannot link to a known craft or pilot. If you're a government, and particularly a defense organization within a government, of *course* you're going to keep an eye on strange ships. What if they're Russian? North Korean? Chinese? Hailing from some homegrown American weirdo's basement?

"What's up?" asked Tasha, pulling into a Starbucks in a small town called Idaho Springs.

"UFOs," I said, flashing my phone's screen at her.

"You gonna look into it?" she asked.

I shrugged, scrunching synthetic down up to my ears. "Maybe," I said, and mostly forgot about it.

I didn't read the article until I returned home that evening, tired, foot-sore. I lay on my couch and called up the page: "Glowing Auras and 'Black Money': The Pentagon's Mysterious UFO Program."

Most of the article—describing a program administered by a Pentagon employee named Luis Elizondo, run by a company owned by magnate Robert Bigelow—fell into the aforementioned "of course" category for me. *Of course* people see objects they can't explain. *Of course* the Defense Department pays good money to investigate them. *Of course* some people think aliens operate them. Then I reached this part:

> The company modified buildings in Las Vegas for the storage of metal alloys and other materials that Mr. Elizondo and program contractors said had been recovered from unidentified aerial phenomena. Researchers also studied people who said they had experienced physical effects from encounters with the objects and examined them for any physiological changes. In addition, researchers spoke to military service members who had reported sightings of strange aircraft.
>
> "We're sort of in the position of what would happen if you gave Leonardo da Vinci a garage-door opener," said Harold E. Puthoff, an engineer who has conducted research on extrasensory perception for the C.I.A. and later worked as a contractor for the program.

People had UFO *parts*? The government was interviewing people perhaps medically influenced by them? Leonardo *da Vinci*?

Once I started looking into these claims—for articles I wrote in the following months for *Wired*—I couldn't stop, even though what I found led me to doubt some of the original reporting. I kept going and came to be writing this book because what intrigued me most was not the UFOs themselves: It was the people obsessed with UFOs.

Nestled among the unhinged conspiracy theorists and those who believe every UFO report like it's biblical resided people I could identify with: logical, dedicated, skeptical. Many of them didn't necessarily "believe" in UFOs at all, and most didn't take the extraterrestrial connotation for granted. Nevertheless, in their spare time—between litigating court cases or teaching middle-schoolers or running planetariums—they scoured declassified government documents, Wayback-Machined old webpages, interviewed witnesses, and wrote up self-aware analyses of their findings.

I undertook this project because I wanted to understand why these people spent so much time on a phenomenon that they weren't even sure *was* a phenomenon—at least not one beyond the human brain.

What I found, when I got to know them, was that we were actually a lot alike in a lot of ways. They sought out mystery in the known world—and then scratched at its surface till it eroded into understanding. They believed people flying high in the government wanted to keep secrets. They craved evidence. They wanted better data. They wanted the truth. They wouldn't—couldn't—stop until they figured it out. That's a lot like the journalistic process.

And in UFOs, they found their ideal problem: a hard and perhaps impossible one that would keep them perpetually busy and engaged, by virtue of its enigmatism, its occlusion, its ever-reliance on first-person accounts, its potential nonexistence.

Looking into UFOs is the investigation that keeps on giving. The conclusion seems to recede into the distance, the end like the vanishing point in a painting: always far away.

II

On the Saturday when the *Times*'s UFO story went live, many other people's screens briefly came alive at the same time as mine. Little lights and light pings tried to pull human attention to cell phones. If the humans received the signal as the senders intended, those humans would feel awe, intrigue, and—most importantly—a desire to click on the story entitled "Real U.F.O.s? Pentagon Unit Tried to Know."

Fingerprints pressed to flat sensors, pressed in passcodes, and phones transported news consumers to a new universe. In this universe, a subject the government had for decades claimed no interest in—identity-less objects in the sky—actually interested the government quite a bit. The military, it seemed, took such sightings—and the craft that created them—seriously. If the stunned readers made it through the story, it would have been easy for them, putting their devices face-down and staring at the wall breathing for a second, to think that the government *all but* admitted that aliens were invading Earth's airspace.

The article didn't say this *exactly*. Instead, it stuck to implication and left pieces scattered on the floor for readers to twist and turn into a picture for themselves. An article that almost simultaneously appeared on *Politico* left room for the same. And if the reader response and the tsunami of coverage that followed are any indication, the inference wasn't lost.

If this inference were true, it would be perhaps the biggest story in human history: We are not alone. Earth is not the only habitable planet out there. It's physically possible to build spacecraft that can cross cosmically significant distances. They're already here. And whoever they are, they cared enough to come to Earth, out of all the planets out there. All those weirdos that we thought were wacko for insisting they'd Seen Something? Maybe they were right all along.

In the *New York Times* story, on screens belonging to both wackos and skeptics, a video began to autoplay. It showed "an encounter between a Navy F/A-18 Super Hornet and an unknown object." A carpeting of clouds floated below the F/A-18, looking just fuzzy and illustrated enough to be from a dream. And in the middle of the infrared heads-up display, the oblong heat signature of some object floated. The fighter jet locked on to it, following automatically as it screamed through the sky.

"This is a fucking drone, bro," said one pilot.

"There's a whole fleet of them," responded the other pilot.

"My gosh," responded his companion, suddenly prudish.

This supposed fleet, outside the frame, was flying against a strong wind, which seemed to impress the Navy men.

"Look at that thing, dude," urged one.

"That's not a [unclear], is it?"

The video faded out a few seconds later.

This video, the article claimed, represented one of many UFO encounters that the Pentagon had investigated as part of an initiative called the Advanced Aerospace Threat Identification Program: AATIP, pronounced *"ay* tip." That program, which ran from 2007 to 2012, had remained secret till now.

Further down in the article, a second video showed a second unidentified object. It had a similarly pill-shaped infrared aura and seemed to zip away from another fighter jet, faster than fast. There were no words accompanying this video.

The existence of AATIP and the government's very interest in UFOs seemed to confirm the spacecrafts' seriousness. The videos brought the message home. And the article told the tale of how this program came to be in the first place, and why the military was investigating a topic it had long denounced as unworthy of study.

As with some other government programs, AATIP was born largely at the behest of one particular politician: Nevada senator Harry Reid. One day, the *Times*'s story goes, Reid was talking to rich constituent and donor Robert Bigelow. Bigelow had made a fortune on his chain of budget hotels. With that money, he'd spun into the aerospace business, building inflatable structures for space. Someday, he hoped, they'd become orbiting hotels. But Bigelow also had a tertiary interest: the paranormal.

He owned a piece of property in Utah called Skinwalker Ranch, a supposed hotbed of spooky happenings. He'd spent years attempting to catch paranormal activity in action so that he could quantify it, measure it with sensors, and make it acceptable to talk about in polite scientific society. To do this research, he'd even formed an official organization, called the National Institute for Discovery Science, and employed scientists full time. The slippery phenomenon, though, always seemed to evade their instruments, and Bigelow shut down the project in 2004. Despite the failure, a Defense Intelligence Agency (DIA) agent had asked to visit Skinwalker Ranch.

Bigelow related this story to Reid, who became curious about what the agent man was up to. So he met with DIA officials and found their subject matter interest was in UFOs. Reid thought the government should be studying those anyway, so with that intelligence community interest and support from two other like-minded senators, he pushed a potential project forward. The three soon initiated the $22 million AATIP. Voila: The Pentagon had a UFO program.

Most of the $22 million—pocket change for the Pentagon and so nearly not noticed—went to Robert Bigelow's company, Bigelow Aerospace. Its people investigated UFO reports, they possessed potential physical material from crashed UFOs, and they examined humans who said UFOs had physically interacted with them. "What was considered science fiction is now science fact," the program's former director wrote in a 2009 Pentagon briefing summary, according to the *Times*.

The origin story the *Times* relayed seemed all but a declaration not just that UFOs are extant, but also that they are extraterrestrial. They have landed, the Pentagon knows about them, and they may have had measurable effects on human beings. It was enough to send chills—fear, awe—down the spinal column. But then, again, there was that third emotion: the doubt.

Da Vinci's garage-door opener? You expect me to just *believe* that? The *Times* authors expected us to believe that they *had* the proof. It's a common practice in journalism—key, really, to the enterprise: *Protect your sources.* And so readers trust that the journalists have done a good job, vetted the material, vetted the people they're talking to, and done the work for the reader. But that's only when journalism is working as it should.

And in this case, it might not have been. Soon after the AATIP story came out, many started to spot holes in the narrative, connections that remained unidentified, questions that demanded answers.

The same day the story came out, for instance, a new website for a company called To the Stars Academy of Arts and Science appeared online. It hosted the same two UFO videos that appeared in the newspaper article. That was just the first of the story's connections to this strange commercial enterprise. On To the Stars's team sat many of the sources in the *Times* article. People like the former director of the Pentagon program, Luis Elizondo; a former high-level intelligence officer, Christopher Mellon; and scientist Harold Puthoff, who did extrasensory perception research for the CIA and said the da Vinci thing. "Their venture aims to raise money for research into UFOs," said the *Times*'s article. On To the Stars's webpage, along with the videos, was a "donate" button for a crowdfunding campaign. The company's CEO was musician Tom DeLonge, former front man of the punk band blink-182.

Suspicions—about the company, its people, and this supposed UFO program's findings—sufficiently raised, I jumped down a wormhole whose boundary I'd been avoiding my whole adult life, especially as a science journalist. UFOs had never interested me much before this point, connected as I was to hard evidence and my own reputation. But having entered said wormhole, I didn't know where, or if, I would pop out. I did know it was a trip I didn't want to miss, because whatever I did or didn't think about UFOs, if someone was going to hint that they had the biggest story of the last several millennia, I wanted to know if it was true—or false—for myself. By midday the next Monday, I was deeper into the ufological Internet than I thought I would ever venture. I spent the first few months researching AATIP, but those months compelled me to try to understand how UFOs are bound up with the cultural, sociological, economic, political, and religious environments of whatever spot they inhabit in spacetime. This book is my attempt to figure that out, for a few key programs, places, people, and times.

I disagree with many of the people here about many things, but the same sentiment animates us: The truth—whatever it is—is out there. Digging it up is both the trick and the thrill.

TWO

THE POLITICS OF
THE UFO CONGRESS

I

It's 7 A.M. when my plane lands in Phoenix on February 14, 2018, touching down on a skim of fallen rain just in time for the opening sessions of a UFO conference. *The* UFO conference, in some people's opinions. Called the annual International UFO Congress, it brings together thousands of people who care about things they can't identify in the sky—people who've seen them, people who want to, people who believe they come from deep space, and people who believe they come from the Deep State. People, in short, who see and want and believe all kinds of things. Having not yet figured out the story of the Pentagon's AATIP, but thinking—foolishly—that I'm close, I attend hoping to learn more about the people who care *most* about that story.

In the rental car, I punch in the address of the We-Ko-Pa Resort and Conference Center and try to plan conversation. While a press badge gives a person a certain license to strike up conversations with strangers, it doesn't magically absolve one of awkwardness. Despite being able to begin with, "I'm a reporter," it's never not weird to walk up to a crush of people you don't know, who might not want you there, who might just want to continue their discussion and sip their $7 cash-bar domestic beers without your intrusion. Which it is—an intrusion into their airspace—no matter what.

Plus, I presume that wherever the conference attendees come from, my origin point is not the same. Presumably, it is primarily their interest in UFOs—whether that interest hinges on belief or skepticism—that brought them here. Whereas it is primarily my interest in *them* that brought *me* here.

One flat, straight road out of Phoenix and toward Fort McDowell becomes another flat, straight road. Mountains—jagged, dry ones with little vegetation taller than a bobcat—press close. Saguaros tee themselves up from the terrain, which seems to be largely made of pebbles.

Every time I see a landscape like this, my mind calls it "alien." I then give that thought the voice-over, "It's so like an exoplanet."

But what my brain means is that this place is unfamiliar. It seems foreign *to me*; therefore, it must actually be strange. Not of this world. But Phoenix is, by its very existence on this planet in the great dusty state of Arizona, terrestrial, not extra. The reaction is irrational. Still, I don't think I'm alone in the impulse to *other* things outside my experience. It's pretty human—a human flaw—to think, "That doesn't belong here," when we have never seen something before.

I turn off these thoughts and turn left into the resort center, parking close to the RVs that, with their stabilizers already down, are set for the duration. Toward the back of the lot sits a big black van. Someone professional wrote "KGRA" in magenta on the side, in a font like tattoos have mated with graffiti. "PARANORMAL RADIO," reads the paint-job's subtitle.

Near the conference center, which is on tribal land, Native American flute music crescendos, siren-songing the crowd to a sign that points the way to registration, which I nevertheless miss, ending up inside the exhibit hall. There, attendees can check out crystals and massage tables and a book about the supposed "secret space program." Which, depending on who you ask, is a shadowy, military-run affair far in advance of NASA's technology (which, you know, weirder things have happened in the black-ops world), or a shadowy, military-run affair far in advance of NASA's technology because aliens and/or their technology have helped.

A few tables down, as if in answer to the questions an alien-abetted space program brings up, sit the Allies of Humanity, whose representatives offer to help humans navigate a world infiltrated by extraterrestrial species. I smile at the hosts and project my desire not to engage. Because even though this would be one of the safest spaces *to* believe in things like that (safer, in fact, to believe than to doubt), I want, deeply, to tell the people I pass that I do not.

In fact and in short, the UFO Congress hosts a whole spectrum of opinions. And before the week's end, the whole thing will seem like a kind of tent revival—but a Big Tent event, the gathering of a group of sects: militant agnostics, spiritual-not-religious types, the church-on-high-holidays cadre, moderates, missionaries, fundamentalists.

This, it turns out, is not an original thought.

In the ballroom, the scenery fully sets the mood. Black banners hang from the wall, the white silhouette of a UFO beaming a spotlight down their fabric. The logos seem to glow: Lights shine from the ground right into the UFO's spotlight, Earth touching the alien and the alien reaching back in some kind of photonic finger-of-God moment. On the three screens at the front of the room, there's an innocuous-looking UFO that has nevertheless beamed down a her-culean bird of prey, which is on fire. It looks like the *Hunger Games* poster. The last set piece is the carpet. And although surely no one replaced the conference center's ground cover just for this meeting, its little circles—disks—float on the monochrome background. What I am saying is that they also look like UFOs.

As the session's start time approaches, the room fills largely with older people, often in couples, nearly all white, who, unlike at least 20 percent of people at any other conference, don't take out laptops. They just wait patiently for the first speaker, T. L. Keller, who will be talking in part about that aforementioned secret space program.

He soon steps to the podium, projecting the confidence of a person who believes they are in possession of knowledge most others have missed. But before he gets into too much detail on that—on the alien spaceship carriers and a military outpost already set up on Mars—he asks the audience a current-events question.

"Who can tell me the significance of December 16?" he asks. This, recall, is the day the *New York Times* published its story about the Pentagon's UFO program. Inquiring about this day—to the UFO Congress—is like asking a class of med students, "What is the purpose of a heart?" They all know the answer. The teacher just wants to hear them say it.

"Tom DeLonge!" someone shouts, which is a simple and metony-mous response. "Made a big announcement!"

"Actually," says Keller, "it was the Department of Defense."

To some in the community, DeLonge has become a kind of prophet, ushering in the truth they've always known to be true. And around him are assembled apostles—Elizondo, Puthoff—who preach the details to the masses. The drip-drip disclosures invite awe, gratitude, ridicule, doubt, jealousy, love, fear. The limbic gamut. But mostly, the recent UFO developments give people what they want: the hope that, soon, the world will be different.

But regardless of how an individual attendee interprets DeLonge and his To the Stars enterprise, it is because of them that this UFO Congress feels more significant than those of the past. Through the outreaching efforts of DeLonge's quorum, the government has seem-ingly validated and vindicated UFO believers, and inspired new ones, by admitting it has videos of question-mark craft. Some, suspecting foul play already, roll their eyes at the December 16 bonanza. But many others smile and clap, all but say, "Amen," forgetting that key admonition from an apostle of a different era: Beware of false prophets.

II

When Cheryl Costa steps to the Congress stage later that morning, she's not there for squishy conspiracy theories, or even to talk about what UFOs are or are not. She's simply there to show numerical analysis on UFO reports. By looking at years of these reports, submitted by everyday people to databases and investigation programs, she set out to get the stats on where reports tend to come from, when, what they describe, and how they change over time. Whatever value or veracity you ascribe to the contents of the reports, there's no disputing a basic statistical analysis of their existence. Ordinary people *do* see things they deem extraordinary.

Costa is practical, not in awe of herself like other speakers. And this salvo encapsulates her attitude: *Who's flying them?* she asks herself.

"I don't care," she responds, also to herself. "I just add them up."

She looks like your favorite geography teacher, who maybe decorates her house in addition to her classroom with tapestries.

"Welcome to the newbies," she says. "Welcome to the true believers. Welcome to the skeptics."

She pauses. Smiles. "And welcome to the Men in Black."

It's mostly a joke.

Before Costa gets into the hard analysis, she softens the audience with her own story of a sighting. She holds her arm out and points her pinky at the ceiling. It was just that big, she says. Just a sphere in the distance. Blue. "Parked there like a rock," she says. It stayed and stayed. And then it left. "When it decided to go, it was gone," Costa says.

"Changed my life," she concludes.

It's a conversion narrative—a single saving moment that tweaked her trajectory. And I think of something historian Greg Eghigian, who's working on a global history of UFOs, once told me: That what grabs people is not that UFOs show up. It's that they go away. It's not that they appear. It's that they disappear. They don't always come back, but they always *could*. You can spend your life searching, seeking.

Having established her personal investment in the topic, Costa moves into the colder, harder facts. To begin, there were more than 121,000 sightings reported in the US from 2001–2015. That sounds like a lot, but the overwhelming majority of sightings don't even get reported. Costa, for instance, keeps her own encounters to herself. Here's who does report reliably, though: Smokers. And people with dogs. They go outside a lot. They know what it looks like outside. They know when something in the sky is off. But they're also unfamiliar enough with that offness to be unnerved.

I look around at the audience. A large portion of them surely believe they've seen UFOs. Maybe, when their UFOs appeared, and then were just gone, those encounters changed their lives, too. Maybe they told their friends, evangelized a little. But maybe they hid their strange light in the sky under a bushel, worried about sounding crazy.

Seeing a UFO, and interpreting it as something extraordinary, seems a little bit like the doomed, lost kind of romantic love. It comes along when you're not looking for it. It amps up your ordinary experience, invigorates your every day. Your neurons fire with something so far outside your normal life that the world seems magical. And then, when it goes away and you return to baseline (after dipping, probably, far below), you start searching for that experience again. But, of course, you can't will it to happen. And it's not the kind of thing you can ever really explain satisfactorily to anyone who wasn't there. It's not even an experience you can really reinhabit yourself, once it's gone.

When Costa's talk is over, I follow a line of older ladies into the bathroom, which is, throughout the conference, 10 degrees colder than everywhere else in We-Ko-Pa.

"Rocket Man" plays through the PA system. The person in the stall next to me hums along: *I'm not the man they think I am at home.*

Oh, no, no, no.

During the break, I wander into the exhibit hall again. Skirting the Allies of Humanity, I continue to the table for a new UFO gathering: the Dulce Base UFO Conference. Dulce is a 2,700-person town in New Mexico, the headquarters of the Jicarilla Apache Nation Reservation. The town sits toward the tail end of the Rocky Mountains, where the mountains are mostly striated red rocks, pushing up at odd angles into flat-topped mesas. In conspiracy world, the location is famous for its alleged alien-human underground installation—Dulce Base—in which aliens and the government were supposed to work together but came into conflict.

When I stop by the booth, the young man staffing it doesn't mention the base, the subterranean sparring, or the military. He doesn't seem concerned, at all, with what might be underground. He's interested, instead, in what's up above.

He's sweet, maybe twenty or twenty-one. He's tired because he doesn't usually get up this early. He also doesn't usually manage a table or conversations with strangers all day. That's his grandma next to him. She founded the festival. He hands me a flyer.

"Everyone in Dulce has seen a UFO," he says.

"Have you?" I ask.

He nods. First came the helicopter. Then, right after, came the blue thing, which broke up into pieces that shot through the sky. Maybe they were together, he says, the helicopter and the thing. The authorities and the aliens, perhaps.

"Maybe it was just SpaceX," he continues, but he doesn't sound like he thinks so.

I tell him I look forward to the festival and move on to the next booth—that of the Earth Defense Headquarters. The director's husband used to work for the government on UFO projects. "He had contact with multiple species," she says. And if I buy one of the DVDs ($20) that has 1,000 pages of text encoded on it, I can learn what he knew about the battles between aliens and humans. The summary on one DVD says it details the "first contact with 'cat' species."

"Is your husband at this conference?" I ask. "Is he still involved with the government?"

"He's in prison," she informs me. For life. For spilling the extraterrestrial secrets, or threatening to.

When I go back into the ballroom, I Google her name + husband + prison to see what comes up. And according to the Internet, he is in jail not for threatening to expose the government's alien coverup but, rather, for murder. He allegedly led a conspiratorial plot to turn the Bay Area's Marin County, just north of the Golden Gate Bridge, into a modern-day Camelot. He planned to physically cut the county off from San Francisco, and protect it with a laser gun mounted on Mount Tamalpais. To help accomplish that, he enlisted two teenagers to help murder and cover up the murder of a local businessman.

Of course, the director said that was just a cover story, disinformation the feds had fed the public so no one would *know* he had ET information. When you can invoke the government's desire to hide information—to be fair, a fair assessment of its machinations, historically, and specifically as they relate to UFO sightings—that takes away anyone's ability to refute the narrative. It's like how some fundamentalist religions say dinosaur fossils live in rock strata not because of evolution and mass extinction but because God wanted to test our faith in creation.

The boy's story I could believe. I believed he saw a blue thing that broke up. But I think there's a slim chance aliens would resemble cats, and a large chance that the director's husband really wanted to turn Marin into a medieval utopia.

III

The next afternoon, a guy named Robbie Graham is giving a talk called "Searching for Truth in All the Wrong Places." Graham, with slim-fit jeans and blazer and a fifties-slick swoop of hair, is younger, hipper, and more put-together than most of the people at the conference. And it is at his talk that a framework for interpreting the conference resolves.

"UFOs is not a subject one chooses to be interested in," says Graham, author of *Silver Screen Saucers*. "It chooses you." And somewhere along his way, he went from an interested but neutralish observer to someone who thought there was a *there* there, flying through the skies.

"It was a painful realization," he says. "That I was a true believer."

My reaction is immediate, and surprises even me. *He seems so normal,* I think. He's smart; he's eloquent; he's with it; he's philosophical. And yet he believes that UFOs exist, sent perhaps from a star many light-years away, staffed perhaps by an alien crew, able to evade conclusive detection for decades. It's absurd. But he helps make sense of himself, imploring the audience to consider that UFO belief is not just *like* religion. It *is* a New Age religion. It's not a science; it's not a hobby; it's a faith system.

Thinking of it that way, it makes sense that reasonable people could, and would, be attracted to it. After all, Christians—a group that includes people who have run countries and particle accelerators and banks too big to fail, as well as cool people who make indie music and writers of great literary works—believe that the son of an all-knowing, all-seeing God, who created the universe *de novo*, came down to Earth and died and came back to life and, in so doing, granted good people eternal salvation. When you state the tenets of *that* belief system, it's actually a lot more absurd than belief in UFOs.

Someone in the audience asks Graham if he's ever had an encounter—close or otherwise—with UFOs or aliens, and how he reacted or thought he would.

He hasn't, he says, which is why he thinks he can be a removed, rational observer of the community. "If I were to actually experience it," he says, "I wouldn't be able to cope with it."

I think of myself, and all my smugness, and I wonder what I would do if I ever really, truly encountered an object that seemed supernatural. Or a member of a sentient species not from Earth, one that no one else saw. Would I be able to find enough evidence to square it with my skepticism? Would I accuse myself of simply hallucinating? Would I hold myself to the same standards to which I hold others? I stare into the magenta stage light that is pointing in my direction. The longer I keep my eyes open at it, the bigger it seems to grow, as if it's coming toward me.

Graham actually edited a book, called *UFOs: Reframing the Debate*, related to this religious interpretation—and, more importantly, about how to interpret and study UFOs in ways that diverge from the fundamentalist tendencies of the current community. The field of ufology, he believes, has stagnated, beset by a large segment that just believes whatever they're told. Anytime a field, or religion, falls too far into its own beliefs and lets go of criticality, it crowds out room for new thought and hard questions. And ufology depends on new evidence to come to new conclusions in a way that traditional religions (which largely left their prophets and revelations in the past) don't, which means critical thinking is key.

At least that's what Graham thinks. True-believe, sure: But don't believe everything to be true. Check it out. Think about it. And bring the power for revelation back into your own brain, rather than leaving it to UFO clergy or the government. It's a kind of liberal Protestant view of ufology.

In the introduction to *Reframing the Debate*, Graham pokes at this. "Today's UFO conferences bear an increasing resemblance to the spectacle of the Megachurch," he writes, "where the cult of personality attracts thousands of believers, all hopeful their prophets can move them just an inch closer to UFO salvation."

Salvation underpins many of the parallels between UFO belief and traditional religion. In most monotheistic religions, there is an omnipotent, omniscient god figure, who offers humans something more and better than what they have here on Earth—an afterlife, redemption, saving grace. This god, and his/her/their promise, provides the humans with something to hope and strive for. There is some sense that this figure guides, or at least watches over, their lives on this planet. And that even if things are bad right now, it will all come out in the eternal wash.

In certain circles of ufology, there exists a pseudo-omnipotent, pseudo-omniscient extraterrestrial species, which also offers humans something more and better than we have here on Earth. Maybe it's unimaginable technology that can take us to the stars or solve the energy crisis. Maybe it's benevolent intervention into world affairs, as they nudge us in the right direction. Maybe it's just the sense that if the aliens can survive and advance for millennia, we can too.

These aliens' longevity and advanced technology provide humans with something to strive and hope for. And even if things seem problematic on Earth right now, someday we mere humans can become like these others. After all, the only material thing that separates Us from Them is their R&D. As Christopher Rutkowski, who conducts the Canadian UFO Survey, wrote for his *Reframing the Debate* essay, "By shifting their omnipotent deity from a spiritual God to a more technological entity, UFO cults place humans at par with their saviours."

UFO belief ebbs and flows, Rutkowski told me, with the tug of social, political, and economic turmoil. "When things get really bad," he says, "you turn to a higher power." If you're more mainstream, maybe you start going back to church. "But for people who are not spiritual in terms of mainstream religion, they will turn to these aliens as omnipotent beings who will have some way of solving those problems," he continues.

Rutkowski titled his essay, the first to appear in the book, "Our Alien, Who Art in Heaven."

A lot of people stopped reading after that. Rutkowski remains a respected public figure in serious ufology, but his religion hypothesis irks the people who probably need to hear it most.

IV

Rutkowski entered UFO world in the mid-seventies, when he was an astronomy student at the University of Manitoba. His department was always getting calls from citizens saying they'd seen something they couldn't explain in the celestial sphere. The professors weren't interested in these UFO reports, especially because the calls often reported things like airplanes, or Venus, or a car-dealership spotlight bouncing off clouds. But Rutkowski's study carrel was right around the corner from the phone. And, wanting to get in good with his superiors, he offered to take the close-encounters calls off their hands.

The task turned into more than brown-nosing, though. Rutkowski became legitimately interested in what people had seen and what they had to say. He started visiting the experiencers—at home, at their farms. "Much to my surprise, they weren't as crazy as my professors had led me to believe," Rutkowski says. "They were just ordinary people who had seen extraordinary things."

Most of the time, he could explain the sightings to people, and he liked the opportunity to teach them about how cool the provable universe is. Still, a few cases left him without answers, as has been the case with every UFO investigation program we know about.

After he left school, he continued trying to be the liaison between the UFO community and everybody else. He wrote papers. He gave presentations. He gained a reputation as "Canada's UFO guy." Starting in 1989, he collected the country's UFO data, from private research groups, law enforcement, and the National Research Council. He turned it into a report: the annual Canadian UFO Survey, now boasting around 20,000 events.

Rutkowski himself has only seen one UFO, in 1977, soon after he started taking those calls. It was just a light, moving in the sky. It didn't look like a plane or a satellite or a star or a planet. And it didn't change his life. He thought, "Huh!" and reported it.

He also forgot about it entirely until he started gathering files for the Canadian survey. "I came across my own report," he says, in the National Archives files.

He laughs, and I ask why he's been so dedicated to this subject—for decades—when he's only seen one, which left no lasting impression.

First, he says he still likes to talk to people about the legitimately weird universe. UFOs haven't cornered the market. Stars are weird! The Big Bang is weird! The size of space is weird! Whatever you're seeing in the sky, chances are, it's weird. Second, he considers himself a UFO pragmatist—objective, but interested. "I'm not out to prove aliens are here," he explains. "I'm not out to prove everybody's crazy. What I'm fascinated with is that we have this persistent phenomenon that continues to manifest despite science's best efforts to sweep it away. To me, it's a human phenomenon that people are experiencing these things." Maybe the humans are experiencing something physical—an actual saucer—but also maybe (probably) not.

The number of pragmatists in ufology, he says, is small. They are far outnumbered by those on the zealous end of the spectrum. "It's good to have an open mind, but not so open your brains fall out," Rutkowski says, quoting astronomer James Oberg. "And there's a lot of gray matter on the sidewalks these days."

Rutkowski had been thinking, for a while, about how all that splatted gray matter makes modern ufology more like a religion than anything else. So when Graham posted in an online group called UFO UpDates, which Rutkowski moderated, asking for essays about out-of-box thinking, Rutkowski offered a proposal for an essay about a different take on current ufology.

In his view, the field has veered in the same direction as the humanities—from modernism to postmodernism to metamodernism. "There's no such thing as a wrong view," he says. "Anything goes, essentially. Ufology people can make claims that are, on the surface, patently absurd, and yet they're being given an audience."

He recalls speaking with a contactee, the term for people who claim to have met extraterrestrials personally. She said she'd been brought aboard a spaceship and told secrets of living a good life in the universe—not cures for cancer, unfortunately, but the usual beatitudes of mainstream religion. Love your neighbor. Treat others as you'd like to be treated. Become vegetarian.

"Do you have any proof whatsoever?" he asked her.

She rolled up her sleeve. "See this arm?" she asked (he did). When she was on the ship, she told him, the aliens had used a special medical device to find a tumor and had then operated on it with a sophisticated medical technique unknown to humans. So sophisticated that it didn't leave a mark on her.

"See, there's no scar whatsoever," she told him. "And that's the proof."

It looked the same, in other words, as an arm that aliens had never touched. "How can scientific methodology deal with something like that," asks Rutkowski, "where science is not even a part of this?"

I think back to another part of the UFO Congress, in which writer and UFO luminary Linda Moulton Howe spent the better part of her talk reading documents related to the "Zeta Reticuli Exchange Program." The program (story goes) began after a live alien crash-landed near Roswell, having come from the planet called Serpo in the Zeta Reticuli star system. Through this singular entity, humans established regular contact—and sent some astronauts to visit—with the extraterrestrials.

The document Howe read was supposedly a transcribed conversation between Ronald Reagan, then-CIA-director William Casey, and a person called only "Caretaker." The latter two are briefing the president on the program, and on the at least five alien species that have visited us. She offered no proof of the transcript's validity.

The woman next to me whispered, "At least five species, wow."

There's no *inherent* problem with unquestioning faith in claims. People are allowed to have whatever kind of faith they want. But it seems more like a stumbling block in ufology, where, by and large, people believe it's possible—if they can just gather and interpret enough information—to understand the phenomenon, or at least prove its existence. It would be like if a Christian said she'd met God, and she meant it literally.

There are, of course, the hardcore skeptics, who devote their efforts to debunking claims of UFOs and close encounters. But they're not really invited to speak at places like the UFO Congress. They update their blogs and populate forums with takedowns. They're like the Richard Dawkinses of ufology, if Richard Dawkins thought there was some compelling evidence that God *might* exist, and thus required immediate attention. Or, perhaps, these Dawkins-esque people hoped to find some.

Outside of the active UFO community, many more people inhabit a middle ground—not the research-oriented kind of Rutkowski, but the "yeah, God might be real, but I'm not going to go to church about it" sort. These people (36 percent of Americans believe UFOs exist, according to a 2012 National Geographic poll) are all over. They're your friends, your aunts, your teachers, you. Just walk into a bar in a town where UFO sightings are common (check Cheryl Costa's stats to get a handle on a good location). Bring up UFOs, flicking at the idea that you won't judge—you just want to know about their experiences. People will talk.

Rutkowsi knows this well. He went to Missouri to watch the solar eclipse in August 2017. On the way, south of Des Moines, he stopped in a little town where he knew people had been seeing so-called spook lights. They're spherical balls of light that supposedly bob and bounce through the air. Rutkowski pulled up to the local bar, which was pretty much the only thing in town anyway. He sat down. He ordered a drink. He said, "Heard about those spook lights?"

"Suddenly," he says, "people were saying, 'Yeah, my friend saw it last week,' 'I saw it the other day,' 'I'll take you where it was.'"

They all, whatever else they believed, now felt that the universe—and, in fact, their very own planet—was a place no one fully understood. There was some mystery and magic left here, after all. There still existed something, even in this tiny town, that felt unfamiliar. And maybe someday, when they weren't expecting it, they'd be lucky enough to experience that feeling firsthand again.

THREE

THE BLACK VAULT VERSUS
THE ROCK STAR

I

A man named John Greenewald identified with that feeling early. For decades, he's dug into UFO facts and fictions in the government's past, and lately he's been working to uncover the truth about AATIP, the Pentagon program revealed in late 2017. In his pursuit of truths fantastical, nefarious, and mundane, he's filed more than 9,000 Freedom of Information Act requests, as of September 2019. They have resulted in 2.2 million pages of released documents about UFOs and other things the ruling bodies don't like to talk about much, many of which had never seen daylight till he asked for them. Today, they're hosted on his website, The Black Vault, whose tagline reads, "Exposing government secrets . . . one page at a time."

Greenewald's trip toward the out-there began in 1996, when he was a teenager surfing the relatively new Internet. After clicking on a search result from Alta Vista, a strange document stared at him from his computer monitor. The bulky hulk of it glowed, shining out at him as he read pages that described an event from twenty years prior—from before he was even born—when military personnel and civilians in the Middle East encountered a phenomenon they couldn't explain.

It had happened on September 19, 1976, the document said. The Imperial Iranian Air Force Command Post had been getting phone calls from suburbanites who'd seen strange lights in the sky. The assistant deputy commander, dubious, reassured these citizens they were just seeing stars. But then, doubting his own reasoning, he called the local air tower. Finally, he went to see for himself. And indeed: There in the sky was a starlike something—but bigger, brighter, than it should have been. Wanting a closer look, the commander called up an F-4 jet from a nearby base.

The jet, though, didn't get the commander what he wanted: As the pilot approached the *whatever*, his comms went out. His instrumentation shut down. He quickly aborted the mission and turned his plane away from whatever it was. All of a sudden, his equipment mysteriously powered up again.

A second F-4 soon took up where the first had left off. Its radar locked on to . . . something. The pings it sent back indicated something the same size as a 707 tanker. As the pilot approached, the object started to move away. Its lights—laid out in a rectangular pattern—strobed from blue to green to red to orange. And then, from *inside* that object emerged a second one. Also lit, also large.

"This second object headed toward the F-4 at a very fast rate of speed," Greenewald read from the report the US Defense Intelligence Agency had released. "The pilot attempted to fire an AIM-9 missile at the object but at that instant his weapons control panel went off and he lost all communications."

Greenewald's eyes passed ravenously over these words as his mind locked on to how like an *X-Files* episode this whole thing was. Maybe this smeary, type-written dispatch was a forgery. How, after all, could it be real? Still, he kept reading, as the pilot dove down and the second UFO tried to intercept him before slipping back inside the first UFO, out of which *another* something emerged. That craft whizzed toward the ground, then seemed "to come to rest gently on the Earth," illuminating the area. The next day, the military helicoptered the F-4 crew to the landing spot. The piece of land looked like nothing had ever happened.

This document *had* to be fake. "This is too cool for the government to give out," Greenewald recalls thinking. But at the bottom of that baby Internet page, its author—anticipating doubt, perhaps urging critical thought—dared readers to find out for themselves. All they had to do was file a Freedom of Information Act (FOIA) request, a boilerplate version of which was included below.

Greenewald, curious, copy-pasted it and sent his query off to the Defense Intelligence Agency, not expecting much. Soon, though, an envelope came in the mail: It contained the same four pages he'd found on the Internet.

The document was real.

So was an article about the event, declassified later, called "Now You See It, Now You Don't," which the Department of Defense published. It lays out this so-called "Iran Incident" like a narrative, beginning with this line: "Sometime in his career, each pilot can expect to encounter strange, unusual happenings which will never be adequately or entirely explained by logic or subsequent investigation."

The whole experience lit a thruster under Greenewald.

After he validated the Iran documents, he thought there must be more legitimate UFO pages online already. But, back in that day, there really weren't. And so he decided to put them there, in what later became The Black Vault.

Greenewald started by simply asking agencies for everything they had about UFOs, which met with enough success to cause a technological challenge: He was a teenager, it was the nineties, and scanners were expensive. Dedicated, he hand-typed each report that came his way, even creating special characters that meant "redacted," "handwritten," and "unreadable." After a few months and three or four hundred pages, he was just starting to go cross-eyed when a businessman from Indonesia (behold, the Internet) wrote to him.

"Do you know what a grant proposal is?" the man asked him.

Greenewald, being a high school student, did not. So the man taught him and then asked Greenewald to write one—describing his need for a scanner and what he would do with it. He must have done okay at his first attempt, because not long after, the man sent him $400.

Today, hundreds of thousands of people visit Greenewald's website every month, reading through the millions of scanned pages, attempting to get the real story. And that was always his goal. "I thought at the time—which was the stupidest decision I ever made which ultimately became the smartest—that the government, out of everybody, would tell me the truth," Greenewald says.

Later, he amended this somewhat: The government might lie, but its *documents* don't lie about themselves. In his decades collecting those documents, Greenewald believes he's found exceeding evidence of federal obfuscation of UFO information. In a book called *Inside The Black Vault: The Government's UFO Secrets Revealed*, he details the government's efforts to convince people that there's nothing to see and to show them that the government itself is not looking, so we shouldn't either. Greenewald has concluded—based on both the papers that exist and those that were destroyed or permanently edited—that neither of those things is true. "That's why I get passionate about it," he says, "because there's so much evidence to prove a cover-up."

Does he think extraterrestrials can be ruled out as an explanation for UFOs? No, he does not. Would he say he *knows* extraterrestrials explain UFOs? No, he would not. Would he say he has enough *evidence* to convince himself extraterrestrials explain UFOs? No. The distinctions between those answers are hard for human minds to make. When you think you can prove an awful lot—but not everything—the brain wants to fill in the blanks, draw the unwarranted connections, and jump to the unjustified conclusions that fit what you already feel to be true. Greenewald calls it the "I Want to Believe" Syndrome.

It went epidemic in 2017, when the *New York Times* released its front-page UFO story. Even though some of the US sees this paper as "fake news," when a story appears within its folds—and, in this case, on the front page—it still carries weight that almost no other news source can. Even ufologists, who generally dislike and mistrust authority and the establishment, still point to the *Times* story's *Times*-ness to indicate its validity and proclaim the topic's acceptance in mainstream culture.

All of that authority has helped the "I Want to Believe" virus spread more widely than usual in the past couple of years.

"I want to believe," as a phrase and a meme, comes from the television series *The X-Files*. It appears at the end of the fourth episode. In the scene, a young Dana Scully sits at a desk, listening to an interview that her FBI partner, Fox Mulder, did while he was hypnotized.

Mulder and Scully have recently formed an investigatory duo, looking into the bewildering happenings for the FBI. These have so far included alleged alien abductions, test pilots who tend to disappear from an experimental Air Force base, and an apparent murderous genetic mutant. Mulder's hypotheses about these happenings, and his sense of their scope, incline toward the paranormal and conspiratorial. Scully, on the other hand, searches for boring terrestrial explanations and attempts to keep Mulder's feet planted on this planet.

But in the universe of this television show, Mulder is—if not right—at least *more* right than Scully. There *is* a vast conspiracy, and its tentacles reach into every aspect of civilization. There *are* aliens on Earth. UFOs and monsters and inexplicable powers of all sorts *do* exist.

Throughout the show's eleven seasons, the team can only access pieces of the whole truth. They pick them up, inspect them at different angles and under varied lighting, and try to fit them together into something that makes sense. They are grasping, always, at evidence whose fullness evades them. That's kind of the state of being in the modern world, even the parts of it divorced from the allegedly paranormal: There is too much information to take in, so much data we could analyze, so many events we could know about, so much background knowledge we could have, so many books we could read, so many topics and programs and people we could be briefed on. We will never truly understand the state of the world and our place in it.

Mulder, nevertheless, keeps trying to understand the stranger parts of his fictional world. And he has a personal investment in figuring it all out: Aliens had allegedly abducted his younger sister years before. And that's why he underwent the hypnosis session to which Scully listens, in episode four, as she fingers the edges of his sister's FBI file. As the tape plays, the images on viewers' televisions switch to footage of Mulder in a church. Light streams through stained glass. He holds a picture of himself and his lost sibling in his hand.

A voice in his head, he tells the hypnotist, promises that his sister is OK, that she'll come back someday.

"Do you believe the voice?" asks the hypnotist.

"I want to believe," replies Mulder.

That sentiment echoes throughout the entire series, often taking the form of a poster in Mulder's office. It features a classic saucer, flying over classic trees, meme-like text at the bottom proclaiming in white all-caps "I WANT TO BELIEVE."

Even if we don't all want to believe in inaudible voices, aliens, abductions, or UFOs, most people can identify with the naked version of the desire. Humans *want* to believe in a lot of things: Some of us want to believe in God. Some want to believe we'll never die, not for*ever*. We want to believe that our partners will love us always. We want to believe that we're good, that our dogs have feelings just like ours, that we can learn from our mistakes.

Given that general tendency, it makes sense that we would also want to believe—even if we don't, exactly—not just that aliens visit us in spaceships but also that there's something *more* here. Our mediocre experience of the world (not as grand as we grew up expecting it to be) is such because the coolest parts are kept from us. Put more negatively, this sentiment means our lives are not completely within our control; unseen forces manipulate them. If this were true, it would—like all those other would-be beliefs—make our lives, if not better, at least more interesting.

If I had a poster on my wall, though, it would say, "I don't want to *have* to believe." I would love to think UFOs are out there and that aliens pilot them. But I am incapable of thinking that unless I *know* it. Still, perhaps knowing is not the point of UFOs. For serious researchers in this field, *trying* to know seems to hold the most appeal of all. Unanswered questions, after all, keep you up at night. They animate you, compel you to crack open that laptop just one more time, letting it light your face blue at 1 A.M. They press you to come up with theories and then test them on your friends. "Hear me out," your sentences begin to begin.

When—if—you find whatever you're seeking, the film of your life slows to fewer frames per second. People I've interviewed have called UFOs various versions of "the ultimate problem to solve." Many of them don't believe UFOs (a term that, denotatively, just means something in the sky that the seer can't understand) were forged in alien furnaces far, far away (although some do), but they do believe these sights are *something*. Maybe they're something in our heads. Maybe they're secret military craft. Misinterpreted planets. Blimps. Wavering stars. Atmospheric phenomena. Swamp gas. Our collective ignorance of all of the above organized into sky lights. But whatever they are or are not, people undoubtedly see things they can't explain, talk about them, write about them, wonder about them.

Sometimes, too, people sing about them. Plenty have penned lyrics about UFOs and aliens: Ella Fitzgerald has "Two Little Men in a Flying Saucer." Queen has "Flash Gordon." Katy Perry and Kanye have "E.T."

But the person belting the loudest these days is Tom DeLonge, the head of To the Stars Academy of Arts and Science, who caused such a stir at the annual UFO Congress. He is also the sturdy-faced former singer and guitarist of the band blink-182. Now, as the head of the To the Stars, his company employs the former director of AATIP and released the same UFO videos the *New York Times* did on the same day. Around that same time, the singer found himself the subject of Greenewald's laser-like inquiry.

II

Greenewald has reservations about this particular front man, but he can't argue that DeLonge is some rock-star-come-lately to UFOs. The singer had started studying them in earnest years before, while his pop-punk band—famous for being naked, profane, and allegro—toured. The band would bounce on the bus for hours, from places like Pocatello, Idaho, to Salt Lake City, Utah, to Denver, Colorado. While they surely filled some of the time by farting on each other, DeLonge also spent solo hours reading conspiracy-minded tomes about UFOs.

It's hard to identify the very first reference DeLonge made to this high-flying interest, but his public acknowledgement dates back to at least 1998, when he leaps from discussing the Warped Tour to pontificating about aliens during an interview.

"I study that stuff, man," he tells the reporter. "UFOs . . . I'll tell you that I think in the next year the US government is going to come out and admit that aliens have visited Earth. The reason I think that is that I listen to this radio show at home which deals with all this stuff. Some guy called up who worked on them for the government and said that he'd been told to start talking about UFOs visiting Earth to start preparing people for the news."

The US government did not, in fact, come out and say that (any of it) in the next year (or in any of the years to follow). But in 1999,

DeLonge took a kind of disclosure upon himself, embedding UFOs into the scatologically titled album *Enema of the State*. On track three, the same heady voice that waxes lyrical about breakups and bad dates also sings, to the same frantic on-beat of power chords and snare drums, about a visitor from another world. The song is called *Aliens Exist*.

> Hey mom, there's something in the backroom
> Hope it's not the creatures from above
> You used to read me stories
> As if my dreams were boring
> We all know conspiracies are dumb
> What if people knew that these were real
> . . .
> Up all night long
> And there's something very wrong
> And I know it must be late
> Been gone since yesterday
> I'm not like you guys
> Twelve majestic lies

The lyrics feel, if not poetic, at least prescient: DeLonge wouldn't receive the UFO Researcher of the Year Award for two more decades, but he was already in pretty deep. He had cleverly hidden a message you'd only understand if you had trodden the conspiracy landscape. The last line of the song, "twelve majestic lies," is an allusion to a widespread, debunked theory about the existence of the "Majestic 12"—an alleged group of scientists, militarists, and politicians who dealt with a crashed spacecraft. The myth is based on a cache of faked documents, which the FBI now hosts on its website, the word "BOGUS" caps-locked and graffitied over the top of them.

DeLonge's bandmates knew about his obsession and understood this wasn't just some song he wanted to sing for fun. The year after the album came out, bandmate Mark Hoppus commented on DeLonge's proclivities, and his gullibility, to *Rolling Stone*. "He's pretty straightforward," says Hoppus. "He hangs out with his girlfriend, and he believes in aliens. Honestly, he believes anything he reads. You could say, 'I read in a magazine that an alien landed in Australia. A doctor found him and did an autopsy—there's footage on the Internet.' And Tom wouldn't even question it. He would take it as gospel and go around telling everybody."

In addition to apparently wanting, hard, to believe, DeLonge might have actually loved aliens *more* than his girlfriend. Another *Rolling Stone* article from later that same year describes DeLonge's first Valentine's day with Jen Jenkins. On that romantic holiday, he and Jenkins were together with roses, candles, lingerie—you know, V-day stuff. DeLonge, though, was tied up with the television.

"There were aliens on TV, but my chick was right there, almost nude, you know?" he told the reporter. "I couldn't decide what to do!" He eventually chose Jenkins, and married her, but you get the gist.

In September 2019, DeLonge filed for divorce.

For a long time, DeLonge went darker on the alien stuff. But he lit up again in 2011, when he blinked on the website Strange Times, a news engine for ufologists and conspiracy theorists. DeLonge started to appear on UFO radio shows like the famed *Coast to Coast*, an hours-long AM broadcast that caters to insomniacs and long-haul truckers and features ads for extremely suspect medical supplements in addition to interviews about the paranormal. Soon, he went mainstream, guesting on *Larry King Live*'s UFO segment as a "musician turned believer."

Around that time, and perhaps not coincidentally, things started to go south for the musician part of DeLonge's identity. By early 2015, the band was beginning to break up.

In the midst of the turmoil, DeLonge took a trip to a region near Area 51, that military test spot where the government supposedly holds and possibly reverse-engineers UFOs (and depending on who you ask, alien bodies or live specimens). It was a place, he believed, where he could experience "The Phenomenon," an all-encompassing term that includes UFOs, aliens, and other paranormal experiences.

"Part of communicating and making contact is shutting your mind down and being able to project your thoughts," he told a reporter for *Paper Magazine*.

Disappointingly, when he and his camping companion tried said thought projection, it didn't work at first. But like with any relationship, timing is everything.

"If anything was going to happen, it would happen at three in morning," DeLonge told his friends, "because that's the time when things like this happen."

At precisely three, DeLonge says he awoke, feeling like his whole body was sparking with static electricity. "It sounded like there were about 20 people there, talking," he recalled. "And instantly my mind goes, *OK, they're at our campsite, they're not here to hurt us, they're talking about shit, but I can't make out what they're saying. But they're working on something.*"

He couldn't move.

He closed his eyes.

And then he woke up. The fire was out. It was six A.M. Where had the time gone?

He had, in UFO parlance, "lost" it (the time, not his mind). It's phrasing the community uses to talk about the gaps in minute measurement that occur when people meet aliens.

The next morning, DeLonge asked his companions about the experiences. Was he alone?

No, turns out.

"They were all around our tent, they were talking," one camper reportedly said. "I told you!"

Right after that, DeLonge had also formed a company called To the Stars. With it, he planned to make music, books, movies, screenprints. "My goal is to create massive audio-visual landscapes that are driven by the concepts and ideas I'm most passionate about," he wrote in the announcement. Those concepts and ideas had to do, of course, with UFOs, ET, and government secrets. Because finding the truth? Hard. Getting "truth" out there via nonthreatening books and movies peddled by a punk rocker? Easy.

Soon, production ramped up. In 2016, To the Stars released a book by Tom DeLonge and *New York Times* best-selling author A. J. Hartley. It was called *Sekret Machines*. It is, according to the book's summary, "a work of . . . fiction?" based "as much as possible" on real places and events. It may also be (not according to the book's summary) an attempted "soft disclosure," the ufological term for gradual revelations about the reality of UFOs and their dealings with our government.

The project began, DeLonge wrote in the book's introduction, with a corporate open house. A friend who had recently retired from a big defense contractor asked if DeLonge would introduce the (unnamed) lead executive to the crowd.

Sure, said DeLonge, if he could have a few minutes of the exec's time. "When the meeting came, I took the bull by the horns, and I pitched him an idea," writes DeLonge, "mostly a benign idea I had for a project that could help the youth lose their cynical views of the Government and the Department of Defense."

Most of DeLonge's 1990s fans probably would have fainted to hear him say "the youth," would have cried in pain to see him as a shill for the feds. But apparently the executive liked this talk, because he allegedly invited DeLonge to come by again. This time, he ushered DeLonge into a secure room, engineered to keep secrets. White-noise sounded in the background so that no one could hear their conversation.

"We cannot be involved in any type of project whatsoever that has this topic associated with it," this executive allegedly told DeLonge, referring to UFOs, "specifically because there's never been any evidence whatsoever that this stuff even exists."

There's evidence that this tale is at least partly true. In June 2019, *The Drive*'s Tyler Rogoway received confirmation from Lockheed Martin that DeLonge had met with members of its "Skunk Works" team. Skunk Works is Lockheed's hush-hush "advanced development" division, which designed the U-2 spy plane, tested out at Area 51. Lockheed's statement read,

> Tom DeLonge reached out to Skunk Works with interest in collaborating on a documentary focused on secret machines and advanced development projects. Multiple members of the Skunk Works team met with DeLonge to explore his vision for the documentary, as we would with any individual or organization interested in telling the story of Skunk Works and the technologies we've developed. We ultimately decided to not move forward with our participation in the documentary.
>
> During this exploration period, DeLonge attended a Skunk Works employee event.

DeLonge won't say exactly what happened in his aerospace meetings. He can't say a lot, really. But he *can* say that, in due time, he assembled a team of people *in the know*. "They all believe I am doing something of value, something worth their time and yours," he wrote in *Sekret Machines*. "All along I was never just another 'conspiracy theorist.'"

OK, most people probably thought, if they thought anything at all, when they heard about this new venture. And, really, most didn't:

The book, which *Talkhouse* called a "Dense-as-Fuck Alien Novel," was reviewed in celebrity magazines and UFO publications. Its influence seemed largest in the existing ufological community—members of which thought it meant that they were *also* not just conspiracy theorists, that DeLonge's book didn't go far enough, that it recycled the same old theories they'd been reading for decades, or that its fictions were just that.

Soon after the book's release, DeLonge quit the band for good. "I can't do everything," he told *Mic*. "I can't tour nine months out of the year with enough time to do the enormity of what I'm setting out to do."

Most articles about DeLonge's new enterprise included implicit eyerolls, hints that the authors didn't take it or him seriously, perhaps thought he'd lost his mind, and wanted to give him the column inches to prove it. But that seemed to change when emails from John Podesta, Hillary Clinton's campaign manager, were leaked to the public.

Podesta has long been interested in UFOs, and has supported efforts to get government documents declassified—which, of course, brought him to DeLonge's attention. The two began to converse, and their conversation came to light when Julian Assange of WikiLeaks let loose a trove of Podesta's emails, including his communications with the paranormally inclined rock star.

In all, more than thirty leaked Podesta emails mention DeLonge's name. Most of the missives between him and the rock star, though, come *from* DeLonge, who wrote ramblings to which Podesta didn't seem to respond. To be fair, Podesta did once inform DeLonge that he had given a CBS interview that included "the topic," the two did set up a call with a "very important General" (who appears to be William McCasland), and they seemed to have an ongoing project together.

But in one note about a supposed deal with *Vice* to produce a docudrama based on *Sekret Machines*, DeLonge wrote that he would love to meet Podesta in person someday soon.

"I hope you get my emails," he wrote toward the end, "and I hope I am not bugging you."

Many took evidence that DeLonge had communicated with a high-level politician as evidence of his ufological validity. But had they given more consideration to Podesta's and staffers' *responses* to DeLonge's missives, they might have strung together a slightly different story. For instance, when DeLonge initially asked if he could set up a meeting to discuss his "sensitive meetings on [his] project," special assistant Eryn Sepp forwarded the note to Podesta along with her own message: "Any desire to set this convo up? Politely hold off? Ask him to send you an email directly?"

Another time, DeLonge offered to catch Podesta up on the ufological goings-on, and Podesta didn't respond at all. DeLonge responded the next day with a meandering, embarrassed email, apologizing for something he hadn't been accused of. It was as if he was emailing a middle-school crush.

"I am an idiot," DeLonge wrote. "I forgot for a brief moment who the hell you are. I apologize for my ridiculous moment of grandeur. Lord—I am honored to be able to work on this with important Men like yourself."

Podesta forwarded this to a campaign staffer and wrote, simply, "Blink 182."

"Thanks! Should I reach out? Would be great to get them involved!!" she wrote back.

"Don't think the band is still together, but I think he would do stuff," Podesta replied.

Another time, DeLonge wrote, "I just announced my project, and the pre-orders of the Novels went up, and kids are mining the Internet asking for any info whatsoever that 'John Podesta' says in the book . . . they already look to you in a leadership role they can trust. And care almost ONLY about your voice in this. That's

HARD to do. Getting young adults to like you, especially if your at your level in DC. Don't lose that. I will brand you much more when this all comes out as a man that the youth can trust and rely on. Not that you care . . . But I do. They do."

"Our secret plan," Podesta wrote in response—not to DeLonge, but to the campaign's communications director.

"Jesus," she wrote back.

Nevertheless, things started to accelerate for DeLonge's phenomenal empire. *Strange Times* was going to be a movie. DeLonge won "UFO Researcher of the Year" from Open Minds, the organization that runs the UFO Congress. And DeLonge kept promising the real revelations were coming soon.

Behind these scenes, DeLonge was busy incorporating the company that would later make it to the front page of the *New York Times*: To the Stars Academy of Arts and Science. After he did this, he really popped onto Greenewald's radar screen.

III

To the Stars Academy entered the spotlight—very literally—around eight months after its incorporation. In October 2017, DeLonge convened a flashy news conference on a giant stage with giant screens. Greenewald watched from home, as DeLonge narrated his unlikely journey in a prerecorded video. Greenewald was rapt, just like he had been reading those first four pages about the Iran Incident so many years before.

"I wanted to shift perception on an extraordinary topic," a DeLonge voiceover said, as the camera showed the musician taking pictures of the White House. In his dealings with spooky officials, he continued, he'd learned some lessons: "One, there are certain things that should never have been secret," he said. "Two, there are secrets that were justifiable at the time but should now be disclosed. Three, there are things that are so unimaginable that certain interests believe they should never, ever be made public. After this, you might even agree."

The introductory video ended, and DeLonge stood on a stage in front of a starry background, a skinny red tie sliding down his chest. He had amassed intelligence-world friends, he said, who had accreted more spooky friends, and together, they had come up with a strategy. Namely: to build a "perpetual funding machine" to "attack" aerospace, science, and entertainment—and also reverse-engineer UFO-type technology, in the form of "an exotic craft with an energy source that can revolutionize the world."

As DeLonge introduced his friends—the crack team of former intelligence officials, former experimental aircraft leaders, and retired DoD authorities—Greenewald felt no skepticism. After all, DeLonge seemed to promise the kind of information he'd been yearning for, FOIAing for, since his adolescence, around the same strange time DeLonge started questing.

Greenewald's mind went supernova when the rock star introduced one previously unknown figure: a soul-patched guy from the Office of the Secretary of Defense, who'd retired just days before. His name was Luis Elizondo.

In a short speech, Elizondo revealed what he'd done at the Pentagon. "By far, the most interesting effort I was involved with was the topic of advanced aerial threats," says Elizondo. "For nearly the last decade, I ran a sensitive aerospace threat identification program focusing on unidentified aerial technologies."

"I was all over that," says Greenewald. "I thought that was so cool."

And now, as part of To the Stars, Elizondo would help release new government footage of unidentified objects, help people submit their own unidentified reports, and help the crowd crunch the data.

Very cool, thought Greenewald.

Even when DeLonge solicited money from the viewers, inviting them to buy shares in To the Stars Academy, Greenewald didn't balk.

"I didn't approach this with 'Hey, this guy's gotta be lying,'" he says, of Elizondo. "It was quite the opposite: This is starting to break."

IV

In fact, this particular break tied back in with some FOIA discoveries Greenewald had made, ones that suggested the government *hadn't* cooled as much to saucers as it pretended and might still be studying them. Back in 2000, Greenewald had discovered an Air Force manual whose fifth chapter dealt with how and where military personnel should report UFOs—this despite the government's repeated claim that it hasn't had any interest in them since its last official investigation program, called Blue Book, ended in 1969.

"Obviously this could be an oversight," Greenewald recalls thinking. "Maybe it was just on the books for years, and they forgot about it."

So he did what he always does: He asked the government for its records, and he watched the manual change from 2000 to 2008. "It was revised multiple times, which meant it was an active publication," he says. And Chapter Five stayed stubbornly there, affirming Air Force interest in knowing who saw what weirdness up there.

Until 2011, that is. At the time, the *Huffington Post* was doing a profile of Greenewald and his wonky website. The reporter asked him, as every reporter does, about the best document he'd ever gotten out of the government. Greenewald, as always, gave him that Air Force manual.

"Just to add legitimacy, I taught him how to download it from the Air Force itself," he says, "to show, 'Hey, this is hiding in plain sight.'"

Being responsible, the reporter called the Pentagon for comment. He waited a couple of days, to no ringback. He called a couple more times. Nothing. Then, late on a Friday night, the reporter called Greenewald.

"Hey, John," the reporter said. "You're not going to believe this."

"He says, 'Chapter Five is gone,'" recalls Greenewald. That ufological section now contained information about . . . hurricanes.

"It shows, number one, their interest, and number two, the lengths they'll go to cover it up," says Greenewald.

Other lengths: The Defense Intelligence Agency and the National Security Agency told Greenewald they did not have any of the original UFO-related records (except for one affidavit), only the secondary redacted versions. Impenetrable black rectangles bar every single iteration. ("I have open FOIAs, so don't kill me if they *eventually* show up," he adds, "but yes, they've told me they can't find them.")

"What's under those redactions?" Greenewald asks. "We'll never know."

Greenewald took that to mean the government perhaps did care about UFO reports, perhaps enough to obscure information about them. So to hear a former defense official, Luis Elizondo, *admit* that he'd run a secret investigation program just a few years ago was intriguing. As always, though, Greenewald needed proof.

After the October announcement, Greenewald started filing FOIA requests about Elizondo's program. He didn't know the program's name, so he asked for "all documents pertaining to the outline, mission statement, objectives, etc of the DOD Aerospace Threat Program."

"Please note," he added. "this may not be the exact, title, but is derived from the testimony of Mr. Luis Elizondo, former DOD employee."

Despite Greenewald's personal interest, though, word of the program and of To the Stars Academy and its desire for funding didn't carry far. A few well-known publications noted the announcement, but for the most part this dropped bombshell didn't explode.

And that, perhaps, is why the crew decided they needed to dream bigger.

Throughout all of this, I hadn't heard about Tom DeLonge's To the Stars Academy announcement. I knew he had a UFO thing going on.

I knew he'd been making promises. I knew he'd alluded to his secret buddies. I'd even—in my more sleep-deprived moments—thought, "Wouldn't it be cool if he were right? Wouldn't it be great if aliens existed and *Tom DeLonge* was the one who knew about it?" After all, I listened to *Take Off Your Pants and Jacket* as much as anyone.

In my mind—as, perhaps, in yours—there's a needling, reptilian part that, when it hears something that would be amazing *if* true, thus imagines that it *is* true. Part of this, for me, is an aftereffect of religion. I grew up strictly Mormon, although I am strictly not religious now. Mormons believe that their worldview—you were a spirit before you came to Earth, you're here to be tested, Jesus visited North America after his death, the true gospel went away, a kid in Upstate New York brought it back in the 1800s, etc.—is the only true worldview. They are *certain* of it. Monthly, congregants traipse to the front of a chapel to proclaim their confidence about this truth into a microphone. In almost all cases, this testimony begins, "I *know* this church is true."

To have grown up with such certainty—I *knew*, too—permanently affected the way I think about the world. Because when I left the church, I realized that what I knew could be wrong.

Now, when I hear about strange things, ones that I know aren't true, a backwards part of me sometimes thinks, "Maybe I'm wrong about how they're wrong."

V

When Greenewald's first FOIA response came back in November 2017, he was stunned. "No records of the kind you described could be identified," it said. A skeptical thought probed into his mind: Maybe the program Elizondo allegedly ran wasn't a UFO investigation project at all. He wanted to give the program, and To the Stars, room to be right, "but things started to fall apart," he says. The *Times* article came out the next month, bringing To the Stars to the mainstream spotlight and giving Greenewald what he *really* needed to do a proper FOIA request: the name of the program. It was called the Advanced Aerospace Threat Identification Program, AATIP. He filed more FOIA requests, and has continued to do so. As of October 2019, some of them remain outstanding, as do several of my own.

While he waited in those early days, Greenewald continued to try to get an interview with To the Stars. But it was to no avail, despite his podcast's—a program called The Black Vault Radio—and website's large, UFO-centric audience. "They were not willing to address any questions ever," he says.

"There's something really fishy here," he thought.

By talking to the Pentagon directly, Greenewald confirmed that Elizondo had worked for the Office of the Secretary of Defense. That seems solid—confirmation that the Department of Defense employed him—but AATIP was a Defense Intelligence Agency program. "[Elizondo] went on the record in an interview where he said something to the effect of 'All roads lead to Rome, and in this case Rome was my desk.'" That, says Greenewald, would not usually be the case if Elizondo worked in a whole different office from AATIP. "That's when I started asking questions," he says. Questions like "Did the Department of Defense really release the videos that the *Times* and To The Stars published?" and "What, really, was AATIP?"

Those questions went, initially, to Pentagon spokesperson Major Audricia Harris. She told Greenewald—as she told me—that the Defense Department had not released the videos in the *Times* article. She didn't address whether or not the videos had been part of AATIP—and she wouldn't confirm that *Elizondo* had been part of AATIP, only that he had worked for the Department of Defense.

"The problem with [the videos], though, is there's no context and they're not original," Greenewald says. "They've been put into an editing system." They don't have beginnings, or endings, or classification/declassification markers. And although To the Stars claimed to have "chain of custody" paperwork validating their origin, declassification, and release, they wouldn't release it. Greenewald says such a chain of custody isn't a thing that exists—for any declassified document.

A break in the case came in April 2019, when Las Vegas journalist George Knapp published a document called DD Form 1910, from the Defense Office of Prepublication and Security Review (DOPSR), asking for three videos—with the same names as the original two, plus one To the Stars had made public later—to be approved for release. This form is just one step in the release process, Harris had said, and does not itself mean the Pentagon has "released" the videos. Nevertheless, Knapp used those same documents to justify the title of his article: "Pentagon Did Release UFO Videos," the polar opposite of the conclusion Greenewald came to—that the Pentagon had not publicly released the videos. And is that not so 2019: for two different people to present the same evidence but with diametrically opposed interpretations.

The form claims the videos related to unmanned aerial vehicles, balloons, and other unmanned aerial systems—not UFOs. In the section asking about future uses for the videos, in terms of presentation or publication, the filer wrote, "Not applicable. Not for publication. Research and analysis ONLY and info sharing with other USG [US government] and industry partners for the purposes of developing a database to help identify, analyze and ultimately defeat UAS threats."

Clearly, these videos *were* published—on To the Stars's site, on the *Times*'s site, and now on countless other sites—breaking the terms of the paperwork, something emails between Elizondo and the review office also confirm. In response to questions about the new evidence, the *Times* reporters told me that the videos "were indeed officially cleared for release by DoD, as proven by a document in our possession." They provided no more information about this document.

Some say that's all semantics: If the videos are the Pentagon's, what difference does it make if the person requesting their release followed exact procedure?

Greenewald, though, says things like this call into question everything else To the Stars says, because it means the company is misrepresenting itself. They also hinder people from getting critical context.

The second video in the *Times* story supposedly shows pilots who took off from the ship USS *Nimitz* to intercept a UFO. But it has appeared on the Internet before, way back in 2007. UFO researcher Isaac Koi, a studious and methodical document-gatherer who (in professional life and under a different name) is a British barrister, noted on his Facebook page that the *Nimitz* video had shown up on the conspiracy theory forum Above Top Secret years before. Koi traced it to the servers of a German film company—one that specialized in creating special effects. Both To the Stars and the Navy have acknowledged that this video was leaked earlier than this latest time it appeared online.

Another strange coincidence has haunted Greenewald. "Forty-eight hours after the date of the *Nimitz* sighting, NASA flew their scramjet engine in the same vicinity," he says. And it wasn't just any flight: This one broke speed records, clocking Mach 9.6.

What are the chances that the world's fastest flight happened at the same time sailors were crying UFO and thinking "aliens"? I asked Matthew Kamlet, NASA Armstrong's public affairs specialist, for details about the test.

The X-43A jet and its launch vehicle took off from Edwards Air Force Base in Southern California, mated to a B-52 bomber like prey in the talons of a bird. The B-52 flew 50 nautical miles off the coast and then dropped the stack, which fell for a second before igniting its engines and rocketing away. It traveled 250 nautical miles while boosting itself, and 600 more miles during engine test and descent. That path would have taken it near the *Nimitz*.

"Of all the gin joints in all the world, it just so happens to coincide in two days?" says Greenewald. "To me, I think there's somewhat a connection there."

If you're *me*, you guess that people got their dates mixed up, or that NASA did some additional tests a couple days later, and the UFO and the jet are the same thing—unidentified because these particular pilots weren't aware, at the time they spotted the object, of the test and perhaps identified later. Or, perhaps, another government entity (American or otherwise) was testing something new and used the test as cover. If you're Greenewald, you entertain another hypothesis, too. "If it's aliens coming down and being intrigued by our stuff, okay," he says. Lately, most researchers don't think the X-43a is related to the UFOs, but the strange coincidence remains.

Regardless, it doesn't appear this video was ever classified, as To the Stars claims. Much of AATIP seems not to have been classified—at all, ever. The government denied a request to restrict its information by turning it into a special access program, and ended it because higher-priority projects took precedence. Further, Harris says, the program dealt with threatening next-generation aviation innovations—weapons that other countries might someday use against the United States. "The purpose of the program was to assess far-term foreign advanced aerospace threats the United States," she told me in 2018, "including anomalous events (such as sightings of aerodynamic vehicles engaged in extreme maneuvers, with unique phenomenology, reported by U.S. Navy pilots or other credible sources)." A different press officer, named Christopher Sherwood, took over the position from Harris in 2019. He stated that AATIP "did pursue research and investigation into unidentified aerial phenomena." That hit the community like a bombshell. So did a revelation from September 2019, when the military confirmed something surprising.

> The U.S. Navy designates the objects contained in the 3 range-incursion videos that are currently being referred to in various media as unidentified aerial phenomena. The 'Unidentified Aerial Phenomena' terminology is used because it provides the basic descriptor for the

> sightings/observations of unauthorized/unidentified air-
> craft/objects that have been observed entering/operating
> in the airspace of various military-controlled training
> ranges—it's any aerial phenomenon that cannot immedi-
> ately be identified. . . . The Navy has not released those
> videos to the general public, nor authorized public release.
> The three videos were/are designated as unclassified.

This statement—which came from Joseph Gradisher, deputy chief
of naval operations for information warfare—perhaps confuses as
much as it might excite (a not-unsurprising result, given that it came
from an office specializing in *information warfare*). Grammatically, it
could mean that they were identified sometime after "immediately."
It could also mean that they were simply unauthorized.

A few days after receiving a statement from Gradisher, Greenewald
got naval information on the videos' dates: The *Nimitz* film was
taken November 14, 2004, and the other two came from January 21,
2015—years after AATIP ended. At the time of publication, there
was no direct proof linking any of these videos to AATIP at all.

Some still conclude these UAPs are UFOs in the extraterrestrial
sense. But in the context of previous assertions, it seems more likely
that the phrasing refers to unidentified foreign craft. More likely than
an investigation into cosmic mysteries is the idea that AATIP was
perhaps not about UFOs in the connotative sense, nor about aliens,
nor about science fiction. As so many things are, it was probably
about war.

An enterprising UFO researcher named Keith Basterfield provided evidence in support of this in May 2018, when he came across
a Defense Intelligence Agency call for proposals, for a program called
the Advanced Aerospace Weapon System Applications program—
which the Pentagon says was another name for AATIP (Elizondo
has contradicted this). This solicitation, on which only Bigelow
Aerospace bid, provides a markedly different picture from the one
To the Stars and the *Times* painted:

The objective of this program is to understand the physics and engineering of these applications as they apply to the foreign threat out to the far term, i.e., from now through the year 2050. Primary focus is on breakthrough technologies and applications that create discontinuities in currently evolving technology trends. The focus is not on extrapolations of current aerospace technology. The proposal shall describe a technical approach which discusses how the breakthrough technologies and applications listed below would be studied and include proposed key personnel that have experience in those areas.

. . . The contractor shall complete advanced aerospace weapon system technical studies in the following areas:

1. lift
2. propulsion
3. control
4. power generation
5. spatial/temporal translation
6. materials
7. configuration, structure
8. signature reduction (optical, infrared, radiofrequency, acoustic)
9. human interface
10. human effects
11. armament (RF and DEW)
12. other peripheral areas in support of (1-11)

It's not inconceivable that the Pentagon used terrestrial language to disguise fringe efforts, slipping its ufological interests under this bland, bureaucratic rug. And given Greenewald's experience documenting similar sleights of paperwork, he's not normally one to simply trust the Department of Defense.

But the evidence is troubling: To the Stars is still being cagey, Greenewald's FOIA requests aren't working. The navy, which supposedly sent reports to AATIP, has said it has no records of communicating about the program with the Office of the Secretary of Defense. The spy version of *Encyclopedia Britannica*—Intellipedia—contained no AATIP entry till after the *Times* story came out, and features no government sources, only public ones. FOIA requests for Elizondo's supposed resignation letter, quoted in the *Times* and expressing distress that the government hasn't paid *enough* attention to UFOs, have come back as nonexistent. When asked to explain why that might be, the *Times* authors told me, "It is not for us to explain their response. We have the resignation letter and have authenticated it."

My request to confirm Elizondo's employment with AATIP went unfulfilled, with the Pentagon saying only that he was a Department of Defense employee. And in June 2019, doubts about Elizondo's actual role rose to a fever pitch, when journalist Keith Kloor published an article in *The Intercept* called "The Media Loves this UFO Expert Who Claims He Worked for an Obscure Pentagon Program. Did He?" He probably did not, it seems. "Mr. Elizondo had no responsibilities with regard to the AATIP program while he worked in OUSDI [the Office of Under Secretary of Defense for Intelligence], up until the time he resigned effective 10/4/2017," Sherwood told Kloor. "We stand by our reporting that AATIP was run by Mr. Elizondo," the *Times* reporters told me.

"I can prove a thousand lies by the military and government, and if they're lying it's another one to tack onto the list," Greenewald says about this whole saga. But he has doubts about all the program's true interest in the unidentified. Sure, UAPs—whatever that means—did play a role. "But, does that mean it was 1 percent of what they did? 10 percent? 100 percent? We need more evidence to call it a bonafide UFO program."

The AATIP story is constantly changing, and by the time this book is published, it will surely have changed again. Maybe we will have discovered that Elizondo did head the program, it was dedicated to UFO studies, and it did find things that the US government can't explain in an earthly way.

It seems much more likely, though, that whatever the public finds out going forward will still be shrouded in clouds, glimpsed as puzzle pieces whose bulges never quite fit into the notches. No matter what information comes out in the future, even if it proves To the Stars has been truthful about everything I've criticized here, one fact remains: Its leadership could have provided clear and direct proof of those claims from day one, and they chose not to. To say, "They were clearly always right," would be pretty epic retconning.

After the *Times*'s article came out, To the Stars was only mildly successful in its crowdfunding, not getting close to the $50 million it needed to reverse-engineer whatever it was going to reverse-engineer. Media and merchandising accounted for 45 percent of its 2018 revenue. So far, as of September 2019, it's released books, music, and a whole lot of T-shirts. And although the team has pushed out a bit of UFO information since—another decontextualized video, a biological report that shows that a tiny mummy is not an alien, a pilot testimonial that's structured to look official but is not, and a research project to analyze physical samples from alleged spacecraft—the opening shout largely attenuated to a whimper. But even outside the company itself, the media empire is going strong: A television show called *Unidentified: Inside America's UFO Investigation* began airing on the History Channel in May 2019, following Elizondo as he looks into cases and witnesses. Right before the show aired, the *New York Times* published a story called "'Wow, What Is That?' Navy Pilots Report Unexplained Flying Objects," featuring stories that also appear in the television show. Strange timing. But now as ever, what warrants a big news story when (and who benefits from it) are all bound up with the entertainment industry. Or, as DeLonge put it in an interview with *Rolling Stone*, "This is the third *New York Times* article that we organized and put together at To the Stars."

In terms of the science-ish stuff, though, To the Stars's biggest resurgence happened in January 2019, more than a year after the initial article. Then, the Defense Intelligence Agency released a list of AATIP-produced reports, sent to a curious Congress.

The reports' titles belong in the basement cabinets with the X-files. They deal with invisibility cloaking, traversable wormholes, stargates, negative energy, gravitational wave communication, antigravity, and the Drake Equation—the classic way to count the number of extraterrestrial civilizations that might be in our galaxy.

But Greenewald (of course) has questions: Did the Pentagon actually *ask* AATIP for these out-there ideas? Or did AATIP, left unchaperoned, produce whatever it wanted? Were these papers just background reading? "If they're just that and just reference, it does not mean the government thinks the Drake Equation is connected to this advanced technology," he says.

Maybe it's in the Defense Department's best interests to let us all sit with this uncertainty. Whether the Pentagon had any hand in creating these question marks or not, whether it studied UFOs or not, if it leaves us with our confusion, we probably won't find the full truth—whether it deals with foreign relations or, you know, really, *really* foreign relations.

"If you muddy the water so much," says Greenewald, "it'll never clear up."

As if to prove the point, just before this book went to print, the Pentagon reversed its prior position. "According to all the official information I have now, when implemented, AATIP did not pursue research and investigation into unidentified aerial phenomena," Pentagon spokesperson Susan Gough told Greenewald in early December. Some said the Pentagon was lying now. Some said it was lying before. Some said spokespeople don't know the real secrets. Everybody thought what they wanted to think.

FOUR

THE GOVERNMENT'S CLOSET

I

The waters, though, have looked muddy since the beginning. In terms of modern UFO sightings, the first happened to a pilot and businessman named Kenneth Arnold. In 1947, he was flying his CallAir A-2 between Chehalis and Yakima, Washington, when he took a detour to search for a downed marine corps aircraft. There was a reward for anyone who could find the plane, and who couldn't use $5,000?

Arnold flew around searching for a while, and accidentally found something else—something much stranger than what he'd actually been looking for. As he watched, rapt, nine objects flew through the air in formation.

That's nothing crazy, really. You'd call it a fleet and go on with your day. But the craft appeared to be traveling much faster than the jets of the time. Arnold allegedly clocked them, as they flew between Mount Rainier and Mount Adams, at significantly more than 1,000 miles per hour. When he landed back on the ground, he—he claimed later—told an *East Oregonian* reporter that the objects skipped like saucers on water, referring to their motion and not their shape. The reporter *wrote*, however, that the craft appeared "saucer-like." That line soon rushed out on the AP wire. The *term* "flying saucer" showed up a day later—the first time of many times to come—when the *Chicago Sun* ran the headline "Supersonic Flying Saucers Sighted by Idaho Pilot." The actual path of the saucer description, from Arnold's mouth to our modern ears, is more complicated: The reporter held fast to the transcription, and, as a National Aviation Reporting Center on Anomalous Phenomena analysis notes, Arnold had plenty of opportunities to correct the record earlier.

"It seems impossible, but there it is," the article ended, quoting Arnold.

Arnold's sighting marks the origin point of modern UFO lore and terminology. His story contains several archetypal characteristics (which it would, of course, itself being the archetype): lights in the sky, spotted by a pilot who *knows* the sky and what should be in it (what insiders call "a reliable observer"), moving fast and with erratic, intelligent-seeming choreography. You could almost swap Arnold with the pilots in the AATIP videos and the military personnel who have come forward since, saying (probably honestly!) that they have seen quick, creepy, inexplicable things up there. Their status as hardened fighter jocks is what lends their stories credibility and unnerves the softer and less experienced rest of us.

For talking about his story, Arnold got more—and different—attention than he would have liked: People didn't believe him. It was only a reflection on the glass, a meteor. He had made it all up. In his own book, *Coming of the Saucers*, Arnold wrote, "I have been subjected to ridicule, much loss of time and money, newspaper notoriety, magazine stories, reflections on my honesty, my character, my business dealings." He was not happy about it, and according to the 1975 book *The UFO Controversy in America*, Arnold said, "If I saw a ten-story building flying through the air, I would never say a word about it." (This statement, though, remains hard to reconcile with the fact that he published his own book, today's edition complete with pulpy cover art showing bathing-suit-clad women holding pictures of outer space up for some saucer pilots to see).

Arnold's sighting, however he felt about it, began an epidemic. And soon, other people around the United States started to see their saucers. The night sky opened up, kicking off a ufological period insiders refer to as a "flap": a period of increased sightings. The term also has the contextual tinge of the word's other definition, "an increased state of agitation." Edward Ruppelt, an air force officer who would go on to be part of governmental UFO investigations, wrote that "in Air Force terminology a 'flap' is a condition, or situation, or state of being of a group of people characterized by an advanced degree of confusion that has not quite yet reached panic proportions." In this case, the people were *not yet* panicking about strange sights in the sky.

If Arnold hadn't said a word, history probably would have nevertheless been set on a similar course. Someone *else's* sighting would likely have catalyzed a similar flap—a year later, maybe two, or five. All events unfold in a cultural medium, after all. And the medium of Arnold's time—colored by the fear of outsiders, fear of invasions, and awe of technology, just like today—was fertile ufological ground. Perhaps, in a world without Arnold's encounter, people would have described "The Phenomenon" differently. Perhaps we wouldn't have the term "flying saucer" at all. Maybe it would have been pancakes or spheres. But *Arnold* and *saucer* are what we've got. So the flap that followed—and, really, all flaps to follow—bear his imprint, however faint.

While we humans like to feel that we choose our own actions autonomously, math and geometry can actually describe their collective nature quite well. And so our waves of UFO sightings tend to take one of two distinct shapes: a sharp peak or a bell curve. The first type is explosive, with lots of people reporting lots of UFOs at once, and then sightings dropping off around the same time. The second has a more tame, tapered onset and a more gradual offset.

Maybe, during either kind of crest, more people really *do* see truly strange things, as could be the case if spaceships or air forces are actually descending. Or maybe the upsurge happens because of what social scientists call "perceptual contagion"—a catching disease, whose sole symptom is that you suddenly notice things that have always existed and interpret them differently because someone else pointed them out. It's like if a friend said to you, "Everyone who wears Abercrombie and Fitch has something to prove." Maybe you'd never noticed anyone in an Abercrombie and Fitch shirt before at all. Now, though, you not only see them but also feel like you know something about them.

Either way, a clear relationship also exists between flaps in the general population and the onset of government programs—a symbiosis that former NASA employee Diana Palmer Hoyt has mapped out. When you view the citizens' sightings and the feds' research side by side, she noted in a thesis paper on the topic, "the dose-response mechanism becomes clear": When the population begins to see saucers, the press begins to say so in the papers. Faced with citizens who expect their leaders to demystify the potentially dangerous mystery, the government has historically *tried* to (not always in good faith). When the flaps were fierce, its agents looked into UFO cases, adding their investigations to the quotidian explanations for the majority of sightings. Citizens are meant to believe that whatever may fly by in the future has a similarly prosaic origin. Don't worry: It's just a weather balloon, a too-twinkly star, Venus, atmospheric physics at play.

When a big flap pops, in other words, codified programs crop up. You can see this happening today, when in April 2019, the navy proclaimed that given the number of unauthorized or unidentified craft that military personnel had encountered recently, it was "updating and formalizing the process by which reports of any such suspected incursions can be made to the cognizant authorities," as *Politico* reported.

Long before that, the first official program came together the year after Arnold's sighting. Like the two programs that would immediately follow, spanning more than two decades of federal effort, this initial effort aimed to soothe—and redirect—the masses, while also more quietly attempting to determine whether these saucers were something the military should worry about. The ethos in general? "Publicly de-bunk and treat the matter lightly," Hoyt noted, "and privately investigate, and take the matter seriously."

The government's first UFO investigation program began the year Scrabble became a game, and the year the United States passed the Marshall Plan, an effort in part to stop the spread of communism in Europe. Also, it was around the time the country began rampant missile testing in New Mexico, thanks in no small part to the German scientists and engineers: After World War II, the government gave German (often from the Nazi party) scientists new identities and fresh lives in America, as part of an initiative called Project Paperclip. It aimed to bring American rocketry to former German heights, while keeping that same achievement from the Soviet Union. With *their* Teutonic know-how, *our* aero-flight program could catch up to the Nazis'—and so to the Russians', who had *also* stolen some scientists from across the border.

Initially called Project Saucer (an obviously bad PR idea), the government quickly renamed its first UFO program Project Sign. It began in January of 1948 and ran for just one year. At the time, rockets like Project Paperclip scientists' were not for spacefaring: They were weapons. But some of these stolen scientists (and their non-Paperclip peers) reasoned that with a little more thrust, the rockets could enter orbit. And with a little more oomph than that, they could *leave* orbit. Despite the less warlordy dream, the country wouldn't send rockets to orbit till the late 1950s. It's interesting that looking out into the universe, we saw our own future and foisted it onto others, already successful.

In the Arnold-era of almost-kind-of-space flight, fears about who might take over or destroy the world flanked the United States. The country had just gotten out of a war, using planet-destroying bombs that the Soviets would also soon possess. The globe felt cold and tenuous. And Project Sign attempted to find out whether the potential conquerors included experimental enemy aircraft or hostile aliens. We're in a similar situation today, with worries about whether the United States will be overtaken by China, about the influence Russia has over our world-leading government. The shadow of international tension looms large, and the language with which Elizondo and others talk about AATIP focuses on the threat UFOs could pose to our nice-boy navy pilots, to our nuclear facilities, to national security, to national preeminence. From this perspective, it's a little like they've managed to capture and redirect our existential fear outward (way outward), while tinging it with awe.

Three months after Arnold's sighting, General Nathan Twining sent a message called "Air Materiel Command Opinion Concerning 'Flying Discs'" to the commanding general of the Army Air Force. The disputed document began:

It is the opinion that

a.   The phenomenon reported is something real and not visionary or fictitious.

b.   There are objects probably approximating the shape of a disc, of such appreciable size as to appear to be as large as man-made aircraft.

c.   There is a possibility that some of the incidents may be caused by natural phenomena, such as meteors.

d.   The reported operating characteristics such as extreme rates of climb, maneuverability (particularly in roll), and action which must be considered evasive when sighted or contacted by friendly aircraft and radar, lend belief to the possibility that some of the objects are controlled either manually, automatically or remotely.

These objects, Twining continued, tended toward the metallic, leaving no trail. They were disc-ish, soundless, and fast. Given a lot of money and development time, the United States could build aircraft with these characteristics, so maybe these UFOs were just UF-OURs, part of a classified project he wasn't privy to. Also possible: another country's. But also possible: They didn't exist at all.

The air force had undertaken low-level, unmandated investigation already, but Twining's memo, some claim, ushered things into officialdom. A few months later, Project Sign was born. It hoovered in UFO reports and sent investigators to determine the hypothetical objects' natures and their threat level.

As the investigations went on, the Sign group split into the two fervent factions, occupying different ends of the ideological spectrum and jockeying for power over the project. Some thought these UFOs weren't really real, and so couldn't be dangerous. This project was thus silly and inconsequential. Another subset of researchers, though, thought the opposite. And these believers soon developed what was later called the Extraterrestrial Hypothesis, a term that has stuck around since and whose meaning remains self-evident.

That leadership polarization—"it's dumb" versus "it's aliens"— has historically posed a problem for Air Force pilots who wanted to submit UFO reports. They never knew to which pole their case would go, or which way that pole's boss was leaning. If one of the naysayers got their hands on it, they might think the pilot was mentally unfit—in general, and especially to be flying planes with guns attached to them. If their report went into the hands of an alien enthusiast, meanwhile, maybe the pilot would become known as *one of them*, and end up a Kenneth-Arnold-type casualty.

II

One of Sign's most notable mysteries happened in summer 1948. Today, we call it the Chiles-Whitted encounter. Back then, people just called it crazy.

It was a nearly clear, moony night. The evening enveloped pilots Clarence S. Chiles and John B. Whitted as they set off in a DC-3 prop plane from Houston to Atlanta. This passenger flight was a fairly novel thing back then, and as the DC-3 passed southwest of Montgomery, Alabama, most of the passengers slept. The one revved rider, though, would be in for a ride. So would the two pilots.

The encounter began when Chiles, the lead, saw a bright light in front of the aircraft. That was nothing too unusual on its face—probably just another version of his own plane. But then the light was getting closer, closer, faster than it should have been traveling. Chiles realized it was traveling too quickly to be any kind of jet he knew. Moving a hand toward his copilot, Chiles tapped him on the shoulder. He pointed at the relentless light, now nearly on them.

Hands back on the yoke, Chiles cranked the plane hard left, and the UFO whizzed past around 700 feet to the right. As it flew by, its belly seemed to glow, emanating a deep-blue light into the deep-blue sky. Orange-red fire blasted from the back. Long, cigar shaped, the craft had no wings but did have two rows of windows, themselves also glowing.

The pilots had just seconds to take it all in before the craft was gone.

They didn't know what they'd just seen. The wakeful passenger, Clarence McKelvie, agreed that something bright and fast had buzzed by his window, but he hadn't been able to decipher more details.

Later, both pilots would draw the alleged ship. Whitted's sketch looks like a missile with windows. Chiles's more resembles a blimp, or more fancifully, the fly for a fishing reel, where the feathers are fire and the fake-insect body is an unknown object with windows for eyes. Their experience fit the classic formula, replicated even today with the AATIP stories: Reliable observers, lights in the sky, *too fast*.

Their encounter made much news, with *The Atlanta Journal* titling its article "Atlanta Pilots Report Wingless Sky Monster."

Boeing's president, William M. Allen, felt confident the monster did not belong to his company. "Not one of our planes," he said to the paper, nor was it any US aircraft he could think of. Within the air force, General George Kenney reportedly said, "The Army hasn't anything like that. I wish we did." That sentiment still echoes today, in statements like that from Commander David Fravor, who witnessed one of the encounters shown in the AATIP videos. "I have no idea what I saw," Fravor told the *Times*. ". . . I want to fly one."

Chiles's and Whitted's sighting may have sunk into history's depths but for a book that soon-to-be Project Sign chief Edward Ruppelt published eight years later, in 1956. In *The Report on Unidentified Flying Objects*, Ruppelt described some Sign staff's response to this supposedly strange night, and all the ones between it and Arnold's: They had written a document called "The Estimate of the Situation."

If you know the military, you know they only give documents that dramatic name when they're very, very serious. "In intelligence, if you have something to say about some vital problem you write a report that is known as an 'Estimate of the Situation,'" wrote Ruppelt. "The situation was the UFOs; the estimate was that they were interplanetary! It was a rather thick document with a black cover printed on legal-sized paper. Stamped across the front were the words TOP SECRET."

We don't know what those in positions above Ruppelt's thought about the situation itself. But they certainly did not like the estimate. Ruppelt claimed they would not endorse the staff's interplanetary conclusion without more evidence. Which seems fair: Not being able to identify objects, even if the technology appears more evolved than humans', still does not lead logically to "ergo, it's aliens."

But there's a bigger problem with the situational estimate than its non sequitur conclusion: We have no proof this document exists at all. "Copies of the report were destroyed, although it is said that a few clandestine copies exist," wrote an investigator named Edward Condon, who would later lead another government-sponsored UFO investigation. "We have not been able to verify the existence of such a report."

Government investigators after Condon also tried to locate it, which suggests they at least believed it *could* exist, and did not entirely think it the fabrication of an officer gone rogue. But it's never been seen again.

In the face of that alleged resistance to its estimate, Project Sign nevertheless continued. By February 1949, it had a new name: "Project Grudge." The research changed tenor and entered what Hoyt called the "'Dark Ages' in the Air Force investigation of UFOs."

"Project Grudge had a two-phase program of UFO annihilation," Ruppelt wrote. "The first phase consisted of explaining every UFO report. The second phase was to tell the public how the Air Force had solved the sightings." While these phases may not have been as explicit as Ruppelt makes them seem, declassified documents do demonstrate that, in spirit if not in letter, part of the government did hope to convince citizens that there was, essentially, nothing to see here.

The Project Grudge team analyzed 244 sightings and explained most of them. But more than 23 percent remained question marks, and the thinking about them resembled some thinking in astronomy. An astronomer may gather data on, say, 100 supernova remnants. She studies them deeply, and she eventually catalogs and explains their characteristics and behavior. When that demographic work is done, she decides that, chances are, the *rest* of the supernova remnants in the sky work approximately the same way—because the laws of the universe are likely the same everywhere. Transposing this to UFOs, if you can explain a fair number of UFO sightings, *chances are* you could explain the others with approximately the same conclusions—if you only had more data, if the human UFO detectors had taken better notes, if those humans were more discerning observers.

But this line of thinking doesn't actually work—not in astronomy, and not in ufology. In astronomy, the few exceptions reveal more about the heavens than the many normals. The supernova that *doesn't* dim as it's supposed to, the star that *pulses* instead of shining, and the pulsar whose pulses *aren't* regular carry the most importance.

Anomalies that remain explanationless but similar to one another may add up to a totally new category of object.

This could be true of UFOs, too. But ufology is a bit different from astronomy, in that a single sighting—just one!—could change everything (not to be dramatic). That sighting would have to be demonstrably, truly not un*identified* but un*identifiable*—and verifiably unpegged to any terrestrial technology. But because of that small-number importance, unless a project can solve *all* the cases, ufologists can't chill. A 100 percent success rate is, of course, impossible, sightings being almost necessarily ephemeral and data-poor, and most of the data coming from human sensory systems, flawed and imprecise no matter how many hours of flight time they have. Which is why, to the unidentifieds' collective presence, the project's final report concluded, "there are sufficient psychological explanations for the reports of unidentified flying objects to provide plausible explanations for reports otherwise not explainable."

And that's not *un*true, either: People do hallucinate. People manipulate. People lie. People misinterpret. The senses and the neurons are fallible, influenceable, untrustworthy, and socially contingent. So *probably* those 23 percent were human errors of all sorts—but what if just one of them wasn't?

That wasn't really the thinking within Grudge, or the rhetorical line it let leak into the public sphere. The most notable article of the time appeared in the *Saturday Evening Post*—a paper known more for its Norman Rockwell covers of chill home life than its discussions of Zeta Reticulans. An article called "What You Can Believe about Flying Saucers" basically repeated the government's conclusions. More trusting of the feds than the average UFO believer, the author largely took officials at their word, while admitting that *others* might express cynicism:

> The hardiness of the [UFO] scare suggests that it might break out again in full bloom at any time, maybe at an embarrassing moment in our international affairs, and, with this thought in mind, I have spent the better part

of two months investigating it. I have had what seemed to be the wholehearted co-operation of the Air Force in Washington and in other parts of the country. I have found that if there is a scrap of bona fide evidence to support the notion that our inventive geniuses or any potential enemy, on this or any other planet, is spewing saucers over America, the Air Force has been unable to locate it.

In reaching this finding, I am necessarily accepting the assurances of the highest officers of the Air Force, and those of its research and development experts that they have nothing concealed up their sleeves. Of course, there are a lot of people, some of them quite sober citizens, who insist that there is "something funny" about the saucer business. These will probably insist that the Air Force is kidding me. But I don't think it is.

The author goes on to cite UFO sightings that have been or could *probably* be explained as extraordinary misinterpretations of ordinary objects, and he debunks them yet again for the reader. But then, strangely, at the end, he asks the reader to report anything funny in the sky to the US Air Force. Further, he tells them exactly how:

While the Air Force finds it difficult to believe that the heavens are populated with inexplicable skimming saucers, diving disks, bounding balls or spooky space ships, even when the testimony comes from such excellent witnesses as pilots Chiles and Whitted, it does want to know about such things.

So, if you're standing out in your back yard or flying your plane some afternoon or evening, and see one of these things in the sky, here is what the Air Force would like to have you do: Before running for the telephone to call your favorite newspaper, take some mental notes on what and where the object is, and what it is doing. . . .

Take a photograph or make a sketch if you can; if not, remember all you can about its appearance and whether it has any protuberances. . . . Note weather and cloud conditions, and observe how it disappears—whether it explodes, fades or vanishes behind clouds. And, of course, if it is obliging enough to crash or shower down any fragments in front of you, by all means secure the pieces—if they seem harmless.

Then sit down and write a letter containing all this information to Technical Intelligence Division, Air Materiel Command Headquarters, Wright-Patterson Air Force Base, Dayton, Ohio. At the same time, maybe you'd better buttress yourself with an affidavit from your clergyman, doctor or banker.

If you've really seen something and can prove it, you may scare the wits out of the United States Air Force, but it will be grateful to you.

Ending the article this way is like saying to your child, "Kidnappers aren't real. Rumors of their existence are greatly exaggerated. They are mostly just your favorite uncle wearing sunglasses and driving a new car! . . . But also if someone asks you to get in a car you don't recognize, call the police immediately." The article, in a lot of ways, embodies Hoyt's take on government programs: Publicly debunk and treat the matter lightly, and privately investigate, and take the matter seriously.

III

At the end of Project Grudge, in 1951, the leaders concluded that unidentified *whatsits* in the sky did not threaten national security and, further, did not bear alien crews. Instead, they identified a few benign saucer causes. According to Ruppelt, who took over during its death rattle, UFOs were

a. Misinterpretation of various conventional objects;
b. A mild form of mass hysteria or "war nerves";
c. Individuals who fabricate such reports to perpetrate a hoax or to seek publicity;
d. Psychopathological persons

The only threat these supposed objects posed, the report concluded, was that "planned release of unusual aerial objects coupled with the release of related psychological propaganda could cause mass hysteria." In other words, if the USSR wanted to destabilize the United States, all it needed to do was throw up some lights that would look weird from the ground, do a little brain manipulation, and then watch as the Americans collectively lost their minds over the coming invasion—thinking not that it was the Reds but that it was the Greens. Given that there was nothing *physical* to fret about, the team recommended that the government cut down on its study of UFO reports, and pass this report to the government's psychological warfare people. *They* might want to know how our enemies could toy with our minds. (Or, because offense is just defense +/- 180 degrees, they might want to know how best to use the tactic themselves).

The year after the government ended Project Grudge, and despite its "nothing to see here" conclusions, it nevertheless upgraded the project to one called Blue Book. The most well-known of the past UFO programs, this one would run for seventeen years under six different directors (including Ruppelt, who actually coined the term "UFO" during the project). Astronomer J. Allen Hynek acted as its scientific advisor, starting out as a skeptic and evolving into an evangelist later in life.

Though formidable, the program did face one big problem immediately: At its onset, *too many* people reported strange things in the sky. The year 1952 marked the beginning of another flap, perhaps catalyzed by an article in *LIFE* magazine called, politely, "Have We Visitors From Outer Space?"

"Who, or what, is aboard?" the article asked. "Where do they come from? Why are they here? What are the intentions of the beings who control them?" These questions, and the ten unsolved cases the article detailed, have a markedly different tenor from the debunky *Post* coverage.

Reports for Blue Book poured in. And that summer came the famous Washington, DC sighting, when radar operators at Washington National Airport saw blips that looked like aircraft. These blips flitted about and hovered near the White House lawn (note: not allowed).

A week after that, a flight crew and the air force both saw strange lights in the same area. Strange objects showed up on radar. These were reliable observers, those were lights in the sky, they were going *too fast*.

Officials would later say the blips, after which jets were dispatched, were due to "anaprop": the anomalous propagation of the radar system's radio waves, which caused a set of mirages to appear, looking physical but actually ethereal. The wonky propagation, investigators concluded, came from a temperature inversion that mixed up the air molecules, causing phantoms to appear on operators' screens. Ufologists have disputed this assessment, maligning it with calls of inadequacy and lack of nuance.

That calm explanation is not, though, what the headlines said. They said "'Saucer' Outran Jet, Pilot Reveals" (*The Washington Post*), "JETS CHASE D.C. SKY GHOSTS" (*New York Daily News*), "AERIAL WHATZITS BUZZ D.C. AGAIN!" (*Washington Daily News*).

With claims coming in from all sides, reports sapped too many intelligence resources, themselves causing a national security risk—just as Grudge had anticipated.

This flap went down when Americans had less faith in the federal shield than they had in earlier years. The Cold War was heating up. Nearly every action an outsider took could be a foreign feint, presaging a blow. Facing those existential threats, people felt uncertain that the government could protect them, whether from nukes or Neptunians. They also doubted that it knew what was going on within its own borders, where, for instance, these UFOs seemed to do whatever they wanted, whenever they wanted, better than our best jets. As pilots come forward now and say the same, those below cruising altitude are similarly concerned. And because the government itself isn't really giving them answers, some look to organizations like To the Stars to be the truth-tellers, because they're the ones who're talking. Kind of.

In response to the international climate and the rising tide of UFO reports, the CIA set up a four-day meeting called the Robertson Panel in 1953, whose findings echo ominously into the present day.

The panel's conclusions, its very existence, and especially its CIA sponsorship remained classified at the time and for many years after. The agency didn't want people to know the government *worried* about their worry about UFO reports. But they did worry, according to declassified copies of the report, which provide a cold-toned assessment of their fears. If foes could use UFOs—real or simply reported—to sow panic among the populace, causing chaos and distrust, that could prime the United States for physical or psychological invasion. And if the Russians could saturate America with UFO sightings, they could launch a weapon and maybe no one would notice because the warning system would be busy chasing ghosts. Even without deliberate foreign malfeasance, if too many people got too amped and called in a panic about Venus, the government would have no resources to sort the MiGs from the chaff.

Watch, the panel also advised, those UFO clubs, civilian investigator groups that had cropped up. Should a flap occur, these groups would have the ears and minds of the people. Keep in mind "the possible use of such groups for subversive purposes." To this day, some ufologists take this surveillance and disinformation suggestion as evidence of their rightness (if there's nothing to worry about, why worry about us?), and of ongoing modern-day watchers.

The panel further reaffirmed some of Grudge's conclusions—most notably that whatever UFOs were or were not, they did not seem to represent a national-security threat. The overload was dangerous, as was the panic, along with the fact that soldiers might see a foreign spycraft and think it was merely one of those UFOs.

But we can fix it, suggested the panel. All they had to do was train people, and do some very public debunking. Agencies could educate employees on how to recognize high-altitude balloons hit by moonlight, fireballs that look like floating orbs, noctilucent clouds that resemble extraterrestrial neural networks.

The debunking should happen in public. Mass media, the panelists said, could also illuminate real UFO stories and their mundane explanations. When people saw something strange, then, they would assume it, like the fireball they saw on that Disney-channel special, was just a terrestrial phenomenon they weren't yet acquainted with.

If you want to know why people read malicious, secret-keeping intent into the Robertson Report, and the investigation programs, you need only read some of the panel's concluding statements, with an ear for their timbre: "The continued emphasis on the reporting of these phenomena does, in these perilous times, result in a threat to the orderly functioning of the protective organs of the body politic. . . . National security agencies [should] take immediate steps to strip the Unidentified Flying Objects of the special status they have been given and the aura of mystery they have unfortunately acquired."

Any time the government decides, behind closed doors, to strip something of *any* quality, that's pretty much a propaganda campaign. And any time the government decides something might disrupt its tightly grasped order, that can read as a license to *impose* order. Given this, it's understandable that the agency didn't want word of their work to get out. It looked bad. It looked like something powerful had taken hold of the American public, and the government not only disliked it but was going to finagle an end to it. If you believe UFOs are a "Phenomenon," you can read the report and see a cover-up campaign.

In keeping the panel secret, the CIA actually sowed the very seeds of distrust they had tried not to plant by keeping it secret in the first place. When word of the Robertson Panel's existence came out years later, the public called for the report's full release. At first, the CIA put out what National Reconnaissance Office historian Gerald Haines called a "sanitized" version. That's all the world had until 1975, when a ufologist named William Spaulding got the whole thing declassified. The UFO-verse was never the same again.

After the Robertson Panel, some members of the Project Blue Book staff felt understandably deflated. A newly revised Air Force Regulation—200-2, about UFO-report investigations—allegedly squeezed some remaining air out of them. When modern ufologists talk about this directive, they mostly focus on two aspects: One, it demands that "Air Force activities must reduce the percentage of unidentified to the minimum." Two, it mandates that the Air Force release information about its investigations—but that that officials stay fairly mum on the unidentified.

Interpreted generously, the first mandate describes a good-faith effort to solve cases more scientifically, and list only *truly* mysterious objects as unidentified. The second is simply an attempt to keep top-secret military aircraft secret. Interpreted less generously, the first is a shadow ban on "UFO" as a classification, encouraging investigators to call something a meteor even if it's probably not. The second is an effort to conceal UFOs from the public.

When the United States found itself in the midst of another flap in 1965, Hynek suggested the military commission an outside, academic study of the phenomenon. And so was born the so-called Condon Committee, led by nuclear physicist Edward Condon of the University of Colorado Boulder. It would reopen Project Blue Book cases and investigate civilian reports, in an effort to get an independent, science-forward opinion on the topic.

In a memo from before the program started, Dean Robert Low wrote,

> Our study would be conducted almost entirely by non-believers who, though they couldn't possibly prove a negative result, could and probably would add an impressive body of thick evidence that there is no reality to the observations. The trick would be, I think, to describe the project so that, to the public, it would appear a totally objective study but, to the scientific community, would present the image of a group of non-believers trying their best to be objective but having an almost zero expectation of finding a saucer.

Ufologists latch on to that "the trick would be" part—and not without reason: This is not the mind-set with which one should begin an objective scientific study. This project, like the government studies, may not have actually been a neutral study but one whose conclusion was foregone: The Condon Committee existed so that it could declare UFOs prosaic and not worth studying. And, of course, maybe UFOs did not deserve scientific study. But when you're pretending to do scientific study, you should at least pretend to deliver—lest you open yourself up, as the Robertson Panel did, to valid criticism that the truth is out there, and you're manipulating it.

Still, the Condon Committee came to fairly mild conclusions: Ufology has not contributed to science, and there is no evidence that UFOs present a threat to national security. As such, we should probably stop spending money on official programs to study them and could, instead, fold investigations into regular military operations.

It didn't shut the door forever, noting that agencies and foundations "ought to be willing to consider UFO research proposals along with the others submitted to them on an open-minded, unprejudiced basis," and that maybe time would bring newer, better evidence.

In response to allegations that the government censored the study of strange things in the sky, Condon's team shook their heads. "We conclude otherwise," they wrote. "We have no evidence of secrecy concerning UFO reports. What has been miscalled secrecy has been no more than an intelligent policy of delay in releasing data so that the public does not become confused by premature publication of incomplete studies of reports."

You can almost hear the people shouting "You're not my dad!" about the paternalism in that statement. People like to see evidence and interpret it themselves—not slurp it filtered from a government they already mistrust. But after this blow, there would be *nothing*, really, to filter. Blue Book toppled here, Condon having bashed in its knees. The program had collected more than 12,000 cases when it ended in 1969, leaving around 700 unidentified. These question-marks—as well as solved UFO cases that researchers believe shouldn't have periods—have helped keep watchers fixated.

And as far as we know, the government never ran an official UFO investigation program again—until around 2007, when it opened up shop at a private contractor's giant office in Las Vegas, a Bigelow Aerospace building with an alien head painted on the side.

The government's past missteps, misdirections, and information campaigns have stamped themselves onto ufological culture today. The government *has* gone into investigations already knowing the answer it wanted. It *has* come to conclusions that don't necessarily account for all the evidence. It has spied on interested parties. It has kept its work quiet. It has downplayed its own interest. And in so doing, it has fostered a culture of conspiracy in ufology. While the actual, documented conspiracies may seem fairly small, what they represent is huge: There *are* things the government doesn't want you to know about UFOs.

Once you, the interested observer, go down that path, the slope is slippery. And the slip is a fun ride if you have a certain mind-set, because it explains not just something about UFOs but also about your life. Opaque, powerful groups hide behind the curtain, pulling the strings, keeping things from you, making decisions for you that you're not even aware of.

You do not completely determine your own experience of the world. To Americans, that's deeply unsettling (if you dislike that your actions can be modeled mathematically, you'll hate that some entity might literally lord over you). Scholar Timothy Melley calls this feeling "agency panic." But while it may unsettle, it is also perhaps a necessary characteristic of a me-me free society. "Conspiracy theory, paranoia, and anxiety about human agency," Melley writes, ". . . are all part of the paradox in which a supposedly individualist culture conserves its individualism by continually imagining it to be in imminent peril." Put simply: If you imagine you might lose something, you'll fight to keep it, whether you should or not, and whether you're actually losing it or not, and whether you had it to begin with. Or not.

If you think your government is controlling your experience of the world, other authorities can fall from their pedestals more easily. While many ufologists hold science in high esteem, seeking to emulate its methods, most think that mainstream scientists just don't *get* it. In our current cultural climate, many people in the United States don't value expertise very much. Being a regular guy, just like you, can sometimes confer more trust and authority than being a person who has studied a topic professionally for decades. Often, people trust the stories they've heard from their friends, about a kid who stopped speaking after getting a vaccine, or how Colorado is getting more snow than ever this winter, or what makes GMO foods toxic. They trust their own experiences and evidence-gathering. There's nothing inherently wrong with doubting authority, not taking conclusions on faith just because someone smart said them, and collecting your own data. In fact, that is what scientists also do. The problems arise when the evidence is shoddy or the conclusions don't follow—which can be hard to see when you're already committed to one side of an issue.

In ufology, researchers often believe the scientific establishment won't look outside narrow paradigms to consider new data, methods, and ways of thinking about the problem. And so—left with an uncaring, tight-lipped government and a blinkered scientific culture—ufologists have taken this -ology into their own hands.

FIVE

THE PATRON SAINT, OR SOMETHING, OF SAUCERS

I

For decades, the civilian who dedicated the most resources to UFO study was Nevada businessman Robert Bigelow. Around the 1990s, he began privately and quietly funding research into "The Phenomenon." This was years before his private aerospace company ran the $22 million federal AATIP. Bigelow had first gotten rich operating the hotel chain Budget Suites of America, designed for long-term inhabitants who need tiny kitchens and monthly rates. And once he'd gathered enough resources, he invested some of them in his true passion: the long-term travels of aliens.

Bigelow, who, through his company, has declined interviews on this topic and did not respond to a request for comment on this book, first became interested in UFOs, he told *60 Minutes* in 2017, after his grandparents told him about an encounter they'd had long ago, when a *something* swooped down from the sky. In the television segment, Bigelow—with a thin, straight-across mustache, aviator-esque sunglasses, and hair practically designed to be tossed by the wind—walks along a gravel road with CBS reporter Lara Logan. They're in a canyon just outside of Las Vegas, in the general area where his grandparents had their encounter. The drapey folds of mountains, speckled with snow, stand tall in the background as he describes the scene.

"It really sped up and came right into their face," he says. As he finishes the sentence, his open hand flies toward Logan's face at high speed, stopping just short of her nose, "and filled up the entire windshield of the car."

It whooshed away just as fast, changing course by 90 degrees and receding into who knows where. This, Logan says, is why Bigelow became obsessed with UFOs. And why an alien logo graces the side of the Bigelow Aerospace building in Las Vegas. It is there that Bigelow's employees work on inflatable space habitats that, someday, may make up modular space hotels.

Bigelow believes humans are not the first to dream of and undertake such space stays: Others have already left their home planets (perhaps riding in levitating inns with kitchenettes). "There has been and is an existing presence," Bigelow says, later in the interview. "An ET presence." He goes on to proclaim that he's spent "millions and millions," more than any other American, studying "this subject."

TV editing makes it hard to know exactly what he meant when he said "this subject," But Bigelow has funded research on UFOs and paranormal activities since at least the early 1990s. He kept his research projects largely quiet, and when they did come up in public, Bigelow's business success seemed to shield him from stigma. After all, he was just another eccentric rich guy. Elon Musk researches telepathy, Jeff Bezos wants a moon colony, Paul Allen sponsored a telescope to search for aliens, and Clive Palmer may want to clone dinosaurs. When you have a high enough bank balance, and have succeeded in being a CEO rather than a guy ranting on the street, people mostly shrug and write colorful magazine profiles about you.

The ufologists Bigelow supported could let a little of his stature rub off on them.

According to a 1999 *Wall Street Journal* article, Bigelow invested $10 million in UFO studies in the 1990s. He'd sponsored a new radio show, called Area 2000, about paranormal topics. Broadcaster Art Bell helmed on the main mic, and segments often streamed in from a slick Las Vegas reporter named George Knapp. Bigelow pulled out eventually, but Area 2000 kept on going, becoming the famed Coast to Coast AM—an hours-long, late-night talk show featuring the fringe, call-ins, and conspiracies.

In 1993, Bigelow suggested that three UFO groups form an official collaboration, under which they could undertake projects that none of them could complete on their own, ones that would produce big data and result in robust analysis. This, the UFO Research Coalition, broke with Bigelow soon after it formed.

Since then, Bigelow has regularly promised scientific projects like those the Coalition intended to undertake. In 1995, he made this desire for rigor more official, founding and funding an organization called the National Institute for Discovery Science (NIDS). It would investigate *phenomena*, from UFOs to the possibility that consciousness survives death (perhaps not coincidentally, his son had passed away three years before)—with real scientists, real instruments, real experiments. But NIDS never delivered in the way many researchers—like a guy named Jack Brewer—hoped.

Brewer, author of the book *The Greys Have Been Framed* and owner of the blog The UFO Trail, believes Bigelow's promises have gone unfulfilled. Data remains elusive; analysis, where it exists, stays opaque. Some of the magnate's projects seem to trespass on people's privacy and organizations' ethics. But ufologists still take Bigelow's money anyway. As Brewer wrote on his blog, "I'm not entirely clear on how a guy can keep bringing spoiled potato salad to the pot luck yet keep getting invited back."

He has a theory, though. "There was a certain prestige in getting that call from Bigelow," he told me. It meant you'd made it as a UFO investigator. And NIDS was supposed to be one of the most prestigious projects.

II

NIDS mostly operated on a remote patch of land in Utah, which people call "Skinwalker Ranch." Bigelow bought the ranch from the Sherman family who, in 1994, had gone there to get away. They purchased the 480-acre ranch in a geographic region called the Uintah Basin. The giant depression formed when tectonic activity warped formerly flat layers of rock. The Shermans' parcel, shaped like a mailbox, was supposed to give them space to start over, alone.

Soon after they moved in, though, they claimed strange things started to happen on their land. They first told their stories to the *Deseret News* in the summer of 1996, right around the time the movie *Independence Day* came out. "For a long time we wondered what we were seeing, if it was something to do with a top-secret project," the patriarch, Terry, told the paper. "I don't know really what to think about it."

Multiple species of UFOs were doing flyovers, according to the Shermans' story, including "a small boxlike craft with a white light, a 40-foot-long object and a huge ship the size of several football fields." Fire-colored portals opened up in the air. Circles of pushed-down grass and depressions a foot deep appeared out of nowhere. Their cows disappeared—or died, mutilated. Then there were the voices. Deep ones, talking in some unknown tongue. Based on a history of UFO activity in the area, the paper reported this part of the basin to be a "hotbed."

The account concluded with gravity. "The Shermans may never know the cause or reason for what they've experienced, but they know one thing," it said. "They just want it to stop."

Robert Bigelow couldn't make it stop, either. But he could take the problem off the Shermans' hands and transform it into his opportunity: With so much alleged activity, both at the ranch and in the surrounding area, this seemed the perfect place to set up scientific shop.

Soon after Bigelow read the article, he offered the Shermans a reported $200,000 for their homestead. They took the deal and moved to a smaller plot of land about 15 miles away. Terry accepted a job from Bigelow—as caretaker of the ranch that had once belonged to him—and then told the *Deseret News* he couldn't talk anymore because he'd signed a nondisclosure agreement.

A month after the deal closed, Bigelow had already built an observation structure, said the *Deseret News*, which kept covering the story despite Sherman's reticence. Bigelow fleshed out his staff, hiring Colm A. Kelleher, who had a doctorate in biochemistry. John B. Alexander—a retired army colonel with a doctorate in education who worked on nonlethal weapons for Los Alamos National Laboratory and whose research starred in the book *The Men Who Stare at Goats*—also joined. Eric W. Davis, carrying a doctorate in astrophysics, and veterinarian George Onet also came aboard. NIDS's advisory board featured a dozen or so members, including Hal Puthoff, a former Scientologist with a PhD in electrical engineering from Stanford and a history of intelligence agency funding; Edgar Mitchell, a former astronaut who founded the Institute of Noetic Sciences to study, among other things, consciousness; and John Schuessler, a former Boeing employee and future director of MUFON (Mutual UFO Network). NIDS attracted serious people with credentials, dubious though some of those credentials may have been.

But as the organization grew, Bigelow kept most people—everyone beyond the property line of the ranch and NIDS's bureaucracy—out in the cold, providing little information about what the scientists were actually doing out there among the sage, scrub brush, and exposed rock strata.

For a while, actually, it doesn't seem like they were doing much at all. In a *Deseret News* report from 1997, Alexander said they had so far failed to capture the phenomena they were seeking. "Definitely no craft," he told the reporter. "No, we haven't had any dog-zappers. . . . Nobody has landed. There has not been anything of significance. . . . We don't have anything to work with."

At some point between then and a 1998 article, though, that stagnation stopped. Bigelow told the paper that the paranormal activity was alive and well ("We wouldn't be there just for the weather," he quipped), and asked residents of the Uintah Basin to report strange sightings to a NIDS hotline.

But while Bigelow wanted the population to play it straight with him, he continued not to pay them the same courtesy. "[NIDS staff] dodged media inquiries, saying public knowledge about their observations would be premature and not in keeping with established scientific methods," said the *Deseret News*. That's wrong, though: Openness is, in fact, a core principle of scientific inquiry.

The eerie quiet unnerved other fringe investigators, "prompting some paranormal researchers to suggest [Bigelow] has a hidden agenda or government connections," said the paper. Bigelow protested the accusation: "People shouldn't be worried that he is part of a covert government group, Bigelow said, but instead should ask themselves why the government, politicians, religious institutions, educators, scientists and the media are not taking UFOs and the possible existence of extraterrestrials more seriously."

Soon enough, all of that would ring wrong.

NIDS operated out of the ranch from 1996–2004, and most of the results remained on lock-down until 2005. Then, deputy administrator Colm Kelleher—with cowriter George Knapp, contributor to Bigelow's Area 2000 radio show—wrote a book called *Hunt for the Skinwalker: Science Confronts the Unexplained at a Remote Ranch in Utah*. In the book, the two authors detail NIDS's scientific setups—cameras, spectrometers, magnetometers, radio-frequency and microwave detectors, and night vision equipment, for instance—as well as the past and present paranormal activity. The narration goes through NIDS's attempts to capture that activity with data-taking equipment. But it was all just that: attempts. They never really gathered significant data.

In the authors' interpretation, the "anomalous phenomena" possessed the personality of the trickster, able to evade their puny manmade tools by anticipating their actions and disappearing when data capture was imminent. Some members of the team dubbed the apparitions' various forms a "sentient, precognitive, non-human intelligence." Scientific studies, after all, demand fancy terminology.

By the time Bigelow shut NIDS down, the authors claimed that the paranormal activity had pretty much halted—spooked, perhaps, by the observers. Which means that at the end of the book, all readers had was what they'd started with. A bunch of stories.

Stories like this one: Two NIDS investigators were keeping watch on the property one night, monitoring a hot spot. One of them had recently meditated because that kind of mental focus sometimes brought out the beasts. The scientists were about to give up and go monitor a new spot when both of them saw a yellow light near the ground.

No, it wasn't a light, one of them proclaimed. It was a tunnel!

And, no, not just a tunnel, he continued: A tunnel out of which a black creature was crawling!

"I only saw that yellow light," the other said doubtfully. "Are you sure?"

"Jesus Christ, of course I'm sure," the seer replied.

At the end of the book, the authors plead with the readers. Try to understand: "What if it happened to you?"

III

Brewer wasn't impressed with the goings-on at Skinwalker Ranch—or the scientists' analysis of them. "I don't think they've given us enough information to know what they did or what they were doing," he says. And that, to him, isn't an anomaly: "I'm not aware of NIDS, or virtually anything to do with Robert Bigelow and his groups of people that have been in his circles for decades, really ever presenting a coherent paper that is anything more than speculation."

Brewer doesn't brook speculation. When he got into ufology, he wasn't interested in the experiences of witnesses. Of a person's anecdotes, another person could only shrug and say, "That's weird." Brewer, instead, wanted to research the people *researching* the experiencers. He is a process guy, a kind of meta-ufologist, compelled by how data and results get disseminated and how individuals' information is treated.

This focus flew in quick after he entered the field. He soon saw that, in the aggregate, ufologists—even the ones with academic acronyms after their names—can't stand up to scrutiny, and often, neither do their claims. "They would avert from being questioned," he says, "and they would complain that they weren't being taken seriously by mainstream scientists and journalists, yet they would obstruct the process of supplying the information that would allow them to be taken seriously."

Brewer spotted some of the biggest trouble orbiting Bigelow's projects, which he thought sometimes exploited the witnesses. He cites a scandal called the Carpenter Affair, whose trail he picked up around 2008. The "Carpenter" part of "Carpenter Affair" refers to John Carpenter, then the director of abduction research for MUFON and a clinical social worker who allegedly practiced hypnosis at his office. In the mid 1990s, it seems that Bigelow paid Carpenter, and Carpenter passed along case files on alleged alien abductions.

Carpenter turned over copies of 140 abduction reports and received, reportedly, $14,000, without informing or asking the people whose lives lived within those reports. Carpenter's wife—an alleged abductee herself—discovered the data breach, Brewer reported, and told her abducted peers. Some of these got mad and got lawyers; some, Brewer reported, stayed in a state of denial. Also, some probably didn't care. But what did Bigelow *do* with those reports? Why did he want them at all?

Who knows. But Bigelow's interest in abduction extends back to his early UFO days. In 1991, Bigelow funded a Roper poll about it, to find out how many Americans might have been snatched. The results indicated that one in fifty, or nearly 4 million, people perhaps had been!

But there's a funny thing about that poll: It never *asked* if the humans had been kidnapped. Humans, see, experience amnesia around their close encounters. Plus, the researchers didn't want those who *did* remember to lie, which they might if they didn't want to sound crazy. So instead of creating straightforward questions ("Have you ever spent a night in an extraterrestrial space hotel y/n"), Bigelow formulated questions whose affirmative answers could *indicate* abduction. Some of the questions the survey asked included:

- Do you remember ever seeing a ghost?
- Do you remember feeling as if you left your body?
- Do you remember seeing a UFO?
- Do you remember waking up paralyzed with a sense of a strange person or presence or something else in the room?
- Do you remember feeling that you were actually flying through the air although you didn't know how or why?
- Do you remember having seen unusual lights or balls of light in a room without knowing what was causing them?

- Do you remember having seen, either as a child or adult, a terrifying figure—which might have been a monster, a witch, a devil, or some other evil figure—in your bedroom or closet or somewhere else?
- Do you remember experiencing a period of time, an hour or more, in which you were lost, but could not remember why or where?
- Do you remember having vivid dreams about UFO?
- Do you remember finding puzzling scars on your body and neither you nor anyone else remembering how you received them or where you got them?

To screen out the overly imaginative portion of the population, the researchers tossed in screener questions, like "Do you remember seeing or hearing the word TRONDANT and knowing it has a special significance to you?" It was a trick: Bigelow had just conjured up the word. If they said yes, researchers tossed their responses out.

When the results of the survey came back, the Bigelow Holding Corporation mailed the report it made to tens of thousands of mental health professionals. *See?* It proclaimed. *Millions.*

Given all of this, when Brewer heard about yet another Bigelow initiative stalking around in 2009, he felt wary. "Wow," he recalls thinking, "how come the guy with the bad potato salad is back?"

Bigelow was back under the auspices of a new company: the awkwardly acronymed Bigelow Aerospace Advanced Space Studies (BAASS), which Bigelow had formed the year NIDS disappeared. The organization "focuses on the identification, evaluation, and acquisition of novel and emerging future technologies worldwide as they specifically relate to spacecraft," according to an archived job posting webpage.

As a BAASS representative, Bigelow approached MUFON, whose volunteers investigate UFO reports. He offered the organization money to create a *new* kind of investigation team: paid experts, with the resources to zip—just like that—to ufological locations the world over, with the best equipment that would collect the hardest data. "Sponsors" were funding this program. Bigelow himself was simply acting as a channel.

MUFON agreed, initiating the "Star Team Impact Project." A recruitment poster shows a black-triangle UFO hovering in front of a too-large moon, near two helicopters, and beside two guys with binoculars. As part of the arrangement, BAASS also sent its own researchers and got access to MUFON's database of UFO reports.

The promising collaboration fell apart after less than a year. At which point people were still wondering, "Where did the money come from?" Then-director James Carrion, disturbed by the sponsorship and Bigelow's behavior, resigned from the organization as a result of the project.

To date, people outside the inner circle haven't confirmed who those sponsors are. But, given that the project's timing overlapped with AATIP, some suspect the government bought this particular potato salad.

By all indications, Bigelow's company was trying to get as many UFO reports as possible, and not just from MUFON. In 2009, Bigelow approached the Canadian UFO Survey's Chris Rutkowski and asked him to pass along cases. In 2010, the FAA requested that "persons wanting to report UFO/unexplained phenomena activity should contact a UFO/unexplained phenomena reporting data collection center, such as Bigelow Aerospace Advanced Space Studies (BAASS)." (Previously, from 2002–2010, the manual had listed NIDS. In both cases, it also suggested the National UFO Reporting Center).

But by 2012, the Defense Department money had disappeared. By 2014 so had the FAA instructions to ring Robert Bigelow about your UFOs. And with that, Bigelow went all quiet on the UFO front, seeming to focus on his inflatable space modules, one of which launched to the International Space Station in 2016.

That same year, he sold Skinwalker Ranch to a mysterious company called Adamantium Real Estate, LLC, for a reported $4.5 million. Adamantium, a fictional substance, exists only in the Marvel Comics universe. It's a suspiciously strong metal alloy not found in nature, its ingredients and their ratios government secrets.

In 2017, Adamantium filed an application to trademark the term "Skinwalker Ranch," stating that its goods and services would include "providing recreation facilities; arranging and conducting special events for social entertainment purposes; entertainment services, namely, storytelling; publishing of printed matter; publishing of electronic publications; Entertainment services, namely, creation, development, production, and distribution of multimedia content, internet content, motion pictures, and television shows." The overlap between paranormal activity, profit, and entertainment rears its head once again.

The identity of the new owner remains obscured from public view, just like the ranch itself, just like Robert Bigelow's research results, just like AATIP details, just like To the Stars's hard data.

Bigelow, as far as anyone knows, isn't directly involved with To the Stars. But the people he's associated with for decades now work for the company. Luis Elizondo acts as its director of global security and special programs. Colm Kelleher is its biotech consultant. Hal Puthoff is the vice president of science and technology. He's also the CEO of EarthTech International, which To the Stars sends its crashed-UFO metals for testing and where Eric Davis works.

Brewer wants answers: about To the Stars, about AATIP, about the ripe potato salad. But many people in the UFO community don't. "Collectively, the UFO community doesn't ask these questions because they're so concerned about having their beliefs validated," he says. "They want somebody to cheer for. And they want a white hat to be proud of."

IV

One half of a duo called the UFO Seekers feels a little angrier about Bigelow, AATIP, and To the Stars. That's part of why, in the description of the YouTube channel, Tim Doyle calls himself "the most hated UFO and Alien Investigator in the field of ufology."

Doyle and Tracy Lee, partners in life and ufological business, started their research after trekking into the desert on dates. They liked being outside—hiking, keeping fit, learning about how the government had taken which land, and what they were getting up to with it. Lee had grown up hunting, knew a lot about exploration and survival.

When they met, Doyle and Lee were living in Bakersfield, a Central California town close to many experimental aerospace sites, like the Mojave Spaceport and Edwards and Vandenberg air force bases. On their deep-dark dates, they'd see strange lights zooming around, witness rocket launches lighting objects to space. When they'd seen enough, they formed UFO Seekers—an operation with more than 200,000 YouTube subscribers and its own logo-laden car and shirts.

On their expeditions, they try to both capture and explain what they see in the sky. And they've seen a lot. Most of it seems startling and so strange it couldn't possibly come from our species. And yet, maintains Doyle, it does. "I wish I had documented every single time Tracy and I went out, captured something, put it on the screen, and it was a plane we didn't know about," he says.

That's why when Doyle saw the initial forward-looking infrared (FLIR) footage from To the Stars and the *New York*

Times, he wasn't impressed. "I would have just thrown the FLIR videos away," he says. He doesn't trust the film; he doesn't trust the interpretation; and he doesn't trust the people who're now taking it public. What they're doing, he says, is entertainment, not investigation.

And that's fine. Actually, he thinks ufology should have both a serious researcher wing and a for-fun wing. The trick is to keep the borders from blurring. It's a separation that cable news fails to observe, that overstated headlines trespass against, and that wholly fabricated stories—which usually tell a certain group exactly what they want to hear—erase entirely. Just like responsible citizens evaluate the bias and quality of evidence in online articles about Iran before believing them, UFO researchers have to take a critical look at the spin and hype shining in their eyes.

"There's going to be an awful moment," he says, "where people wake up one day and realize there were these professional wrestlers in the UFO community."

In UFO Seekers' years trekking into the desert, they've only seen one thing they couldn't explain: in the Sierra Nevada mountains, looking into military airspace. They had a wraparound view, 10,000 feet up, with Mount Whitney poking into the sky in the distance.

That night, Lee took more than a thousand pictures of low-flying jets. At the time, the jets were all they noticed. But after they took the pictures home, Lee caught sight of a circular object—a silver ball with a single cycloptic red eye that seemed to have a sonic-boom cloud around it. Two pictures—taken over less than a second—showed it flying at around a 50-degree angle into a canyon, then changing shape. "We never heard anything," says Doyle. "We never saw anything."

"If you were looking for aliens visiting earth in our work," he says, "that would be the closest thing we've captured."

When he posted video of it on YouTube, a commenter suggested it was likely an experimental drone project. Maybe. Probably. Who knows.

Another time, Lee and Doyle camped out in the Kern River Canyon, a majestic divot full of fractured white rocks and more water than you normally see in this part of the state. Suddenly, they heard a noise. Looking south toward Los Angeles, they saw an orange fireball scream across the sky. It quickly morphed into a long streak, now made of hundreds of objects that each seemed to be burning up. *What the fuck*, they thought. And then they got to researching.

The local meteorologist said it was a falling star, a piece of outer space dropping down to Earth. Later, though, Doyle discovered this object had come *from* Earth to begin with: It was an old communications satellite, part of the Iridium constellation—the same kind whose flares, from orbit, people like me have often mistaken for spaceships—deorbiting down through the atmosphere. It was, finally, coming back home.

V

Soon after AATIP and Bigelow's role in it came to light, I took a last-minute road trip to Skinwalker Ranch, dragging my sister along for company. Bigelow wouldn't talk, his researchers weren't there, and he didn't even own the property anymore. But it drew me in, even though I knew I'd almost certainly see nothing and even though I had no investigatory reason to go there. Really, I wanted to know if it felt different from other places, like Area 51 does. I wanted to see for myself if it was special. Plus, the (outsider-run) website www.skinwalkerranch.org meticulously details the places a person can stand next to their car and wait for high strangeness to happen, the closest you can get to the hallowed property without trespassing. Not that you'd have a great shot at getting far: On the main road, there's a big gate, menacing stop signs, and creepy cameras.

And so it was that on a sharply cold, clear night, we found ourselves huffing hot air into our hands on a gravel pull off (labeled simply, on the website, "Viewing Area") adjacent to Hilltop Road, looking down on the place that had so scared the Shermans, that had so eluded investigators, that would soon be a hotbed of entertainment.

We were not the first to come here: mostly smoked cigarette butts looked like Pick-Up Sticks on the ground. Crushed beer cans hid behind rocks. I turned on an audio version of *Hunt for the Skinwalker*, and the literally unbelievable accounts poured into our ears as we kept watch for whatever. The narrator's voice was the only sound except for the occasional pickup that would swing too fast by us, all going the same direction, down the same lonely road. I didn't feel much of anything, really, except cold and silly.

And I didn't expect anything to happen, of course. I didn't believe the stories. I didn't think what went on out here was anything stranger than predator-prey relationships and paranoia. But, also, I am a human: There was a part of me that thought, "What if? What if it happened to us?"

If it had, we would have been even worse off than the NIDS investigators: We possessed no measuring devices, no instruments more sophisticated than old-model iPhone cameras. If tunnels or lights or Skinwalkers in any form had appeared and then disappeared, we would have collected only the experience.

I don't think I would have believed my own eyes, anyway.

On the way home, we stopped by Dinosaur National Monument, where the Department of the Interior has constructed a three-walled building against a rock-face of fossils. The fossils, here, form the fourth wall. When we arrived, it was early, foggy, chilled. We were the only ones there besides the park ranger. She had opened the building just for us, just so we could see the remnants of this reptilian civilization—ripped into pieces or stilled in tableau—captured in all their spiny strangeness. We'll never really know what their lives were like, or have anything but this fractured evidence of their existence. Since the beginning, the world has been so full of so much that will always evade our full understanding.

"Let me know if you have any questions," said the ranger.

SIX

THE MUTUALLY SUPPORTIVE
MUTUAL UFO NETWORK

I

On a Saturday afternoon in January, the parking lot and residential streets near Englewood High School are filled with slushy snow, piled into never-melting mounds at the margins. Kids pack the institution's hallways, just trying to have their volleyball tournaments. By them, though, stands a sign with a giant green arrow and the logo for the Mutual UFO Network (MUFON) on its bright white front. The arrow points to a high-volume classroom, where acoustic panels line the walls and one must wonder what all these teenagers in their tiny shorts think of the olds who, after stocking up on Fritos and soda from the sports-team concession stand, head to their fringe gathering.

A woman named Katie Griboski sits sideways in the second row, surveying the saturating room with obvious pleasure. "We have three new potential field investigators," she notes, referring to the volunteer cadre that digs into the UFO reports that anyone can file online. Griboski is one of the state's (one of the country's) premier investigators, at least according to Colorado director Doug Wilson, who's leading this session.

When the meeting finally begins, Wilson—jovial, wearing a collared MUFON polo—steps to the front of the room to give a retrospective on 2018, a year when field investigators dug into 187 UFO cases that people submitted. The state's sleuths spoke to each of those witnesses, trying to figure out what, exactly, they saw in the sky.

Which, mostly, wasn't aliens: Wilson's researchers successfully identified 82 percent of the sightings. Around 16 percent were simply natural phenomena, misperceived. Forty-five percent were human technology, misattributed. Hoaxes accounted for around 10 percent of cases. Seven had insufficient data to reach a conclusion (submissions like "I saw a light," with no location, date, or traits), and some viewers sent cases "for information only." After all of that, just 18 percent kept the label "unidentified," going into the MUFON database as official "unknowns."

Wilson admonishes the crowd about what that word means.

"It doesn't say 'spaceship from Venus,'" he says.

Later in the meeting, Wilson underscores this point. "Raise your hand if you believe in UFOs," Wilson commands.

Hands, of course, go up.

He gives the second option: "If you're not sure."

This time, fewer hands raise. Griboski, who prides herself on her skepticism, is among the ambivalent. Wilson points at her. *She* is a star investigator, *and yet* she is unsure. In fact, she is a star investigator *because* she's unsure.

"There's a third category," Wilson says, looking down like he's guilty by proxy. "Those who *know*. Who's in that?"

People eye at each other uneasily and then lift their arms.

"I need to talk to you," scolds Griboski. What she means is that there is no "knowing."

"Talking to you," though, is the essence of local MUFON groups. Sure, the organization as a whole has amassed one of the world's largest databases of UFO reports. Sure, it sends people to study them. Of course, it distributes a newsletter, trains investigators, holds conferences, compiles statistics, and charges quite a bit of money for everything. But at meetings like this, in high schools and libraries and spiritual centers across the world, all that data and bureaucracy seem secondary. Mostly, community MUFON meetings draw a safe circle. Inside it, you can talk about whatever kind of UFO experience you want, for however long you want, and people generally won't judge it or you for anything other than what it is or you are: Your experience, yourself. Listeners won't walk away, freaked out or embarrassed, no matter how out-there your story sounds. So, for example, when a lone attendee at today's meeting—wearing a ski bib over a fleece over a sweatshirt—raises his hand to say he watches the runway at Denver International Airport slide back, quick as a blink, to let a spacecraft out, people nod. Even if they don't believe that particular story, they *get* it. It reminds me of stepping into a gay bar when you're gay: Your cells seem to relax, because every neon-lit person knows you in a way straight people can't, even if you've never met, even if you never will, even if you're very different. And for once, you can fully let go of the fear of being the self that simmers beneath the calm banality you present to the rest of the world.

"It's important for us to get together," Wilson tells the crowd, "and to talk about a subject that for a long time I couldn't even talk about at home."

You can almost hear the "amens" from the audience.

MUFON is maybe the most centralized group of its sort, but it isn't alone. It swims in a sea of other acronyms, which have appeared and disappeared over the years. Early on, in 1952, the Aerial Phenomena Research Organization (APRO) started up. Two years later, Civilian Saucer Intelligence (CSI) formed. Then came NICAP—the National Investigations Committee on Aerial Phenomena—founded in 1956 in part by an inventor who also claimed to have discovered an anti-gravity force. It boasted prominent affiliates, like Marine Corps Major Donald Keyhoe, who went on to pen the book *The Flying Saucers Are Real*, and former CIA director Roscoe Hillenkoetter. While it attempted to hew to scientific methods, the organization eventually dissolved due to financial mismanagement and infighting, although many of its files remain available. Humans.

In the 1970s, UFOIL—the UFO Investigators' League—flew in, as did the Fund for UFO Research (FUFOR), both now extinct. That decade also saw the rise of the National UFO Reporting Center (NUFORC), currently run from a former missile silo in Washington. Since its inception in 1974, the group has maintained a hotline through which people can submit reports.

MUFON appeared in the middle of all this, a schism from that earliest group. One of APRO's Midwestern members—a guy named Walt Andrus—formed the Midwest UFO Network, taking a large fraction of APRO's adherents with him and eventually changing the "Midwest" into the "Mutual" of MUFON. As the other groups withered and died, MUFON scooped up their members.

MUFON outwardly desires scientific methods and skepticism. In practice, though, it leans more toward belief—in the existence of UFOs, and in their otherworldliness—than it proclaims. According to an article from *The Skeptical Inquirer*, by noted skeptic and author of the blog Bad UFOs Robert Sheaffer, that pseudoscientific attitude has been there since inception. Andrus, he wrote, "was barely able to tolerate skepticism in any form," and excommunicated members who expressed too much doubt.

If you bumble around on the Internet, you'll find plenty of similar, more modern complaints about MUFON's alleged lack of rigor, its endorsement of junk science and conspiracy, and its corruption. For example, in 2017, MUFON's Pennsylvania director wrote a racist Facebook post, asserting (among other claims) that "Everything this world is was created by Europeans and Americans." Other words in the post: "F'ing blacks."

Racism and white supremacy are unacceptable any time—but they are also philosophically inconsistent with the ethos of an organization that thinks not just beyond our skin colors and countries of origin but also beyond our species and planet. In the scandal's wake, MUFON's director decried the age of social media and called for both-sides-ism. Top officials (and presumably some lower-downers) resigned in protest.

Yet many ufologists still congregate around MUFON, despite knowing of its problems, in the same way that Catholics who don't like the church's expenditures, pedophilia, or stance on abortion nevertheless attend mass. Because of belief, connection, tradition, compartmentalization, dissociation. Because . . . humans.

To understand what tractor-beamed Wilson to MUFON, I arrange to meet him at a coffeeshop in the Baker neighborhood of Denver, a gentrified place with punk bookstores near restaurants with cheese plates labeled $$$ on Yelp.

"Now I remember why I don't come to this part of town," Wilson says, sitting down with an iced tea. "There's no parking."

Wilson is a sturdy, bearded guy. He leans against the railing next to our table and pulls out a laminated badge. "We're supposed to make sure we identify ourselves," he says, "and that way, you know who I am."

"I appreciate that," I say, checking out his card, which boasts acronyms that I don't know, symbols and letters indicating the special subteams he's part of. Every group has to have their own language. The incomprehensible words unite those within the circle, partition off those without, and let people feel very special when someone lifts them over the wall.

He puts his card back in his wallet. I've picked a spot at the back of the Metropolis Cafe, a place full of people plugging away on laptops and definitely not talking about UFOs. The particulate smell of espresso dust permeates the air and condenses on my sweater as Wilson begins his origin story: Before he directed MUFON, he had researched UFOs for thirty years as a free agent.

After his seventeen-year marriage ended, Wilson become kind of obsessed with Area 51, making the twelve-hour trek to the Nevada desert all the time. Because he could. Because belief, connection, tradition, compartmentalization, dissociation. Somehow, in between all that travel, he met someone new back in Denver. "She told me early on, 'I'll never try to change what you do,'" he says.

Nevertheless, she saw him getting older and worried about him running around in the wilderness, hunting for things (earthly or celestial) that entities (human or alien) didn't want him to find. She urged him to join a UFO group. And the group that existed was MUFON. Six months after he signed up, he became Colorado's chief investigator. And soon after that, he became state director.

He's glad for the gig, but the bag is mixed. When Wilson was on his own, he only looked into cases he thought were credible, whose results might be worthwhile. That's not the MUFON way, he says, launching into his quibbles with the organization: They don't exercise the same discernment or exhibit the same sense of priority as he did on his own. "At MUFON, we are required to investigate every case that comes through," he says. "MUFON and I disagree on that tactic, but I work for them, so I do what they tell me." Wilson feels bad about assigning bunk cases to his field investigators, so he keeps the worst for himself. Which means he hasn't had a decent case in a long time. Nevertheless, he told me later, "I certainly understand why MUFON wishes to give every case 'due diligence,' both for the thoroughness of the data and for the integrity of MUFON."

He glances at the table next to us, where two women are preparing a slide deck. They're young but trouser-cut, professional. They glance back with increasing frequency as Wilson goes on, describing what happens when you file a report with MUFON, which involves clicking through various multiple-choice options (blinking lights or stable ones? Triangle or chevron shape? Occupants or no or N/A?) and writing a description.

Within six hours of pushing "submit," a regional high-up has read your report of hovering lights, crashed craft, or abduction. That person assigns the case to a field investigator. If it's a Type I encounter (lights in the sky), that investigator has to reach you within forty-eight hours. If it's of Type II (physical artifacts or effects), they have thirty-six hours. And if there's an actual alleged being—an entity, as insiders like to say—investigators will be on it within twenty-four hours.

While the reporting form gives plenty of room and options for details, lots of the reports that come in lack substance: They're missing basics like times or dates or directions, or they give such sparse description that there's hardly a way to follow up. The other day, for instance, Wilson got one that simply said, in the box for a short description, "numerous entities and spacecraft." In the section with more room for verbiage, this person said, "short description speaks for itself."

"My first reaction to that is, 'It goes in the trash can,'" says Wilson. "MUFON says, 'Investigate it.'"

The investigator's job, at first, is to sort the seer's jumbled thoughts and check-marked sighting characteristics into something sensible. Then comes the real work of figuring out what that person actually *saw*. Sometimes, cases are easy. It's Venus, wavering in the atmosphere; it's a jet, coming straight at you so that it seems to be hovering; it's ball lightning, looking like a spooky orb. Others require more detangling. Investigators call local airports to check on flight paths. They call military bases to see whether tests were taking place. They look at launch histories. They go into the archival weather records to see what tricks atmospheric layers might have played. They look at previous cases in the area, to see if anyone spotted something similar—in the past or, if they're lucky, on the same day. The investigator logs their work, keeping track of sources, and creates a final report that sometimes spans up to 100 pages.

THEY ARE ALREADY HERE

"It's not unlike doing a police investigation or a criminal investigation," Wilson says.

So far, Wilson's outlining has all been procedural, his own language like that of a policeperson who says "vehicle" instead of "car." It's a diction of depersonalization and generalities. But suddenly his demeanor changes. "I'll tell you a little bit about a very exciting case," he says. ". . . As soon as I saw it, I thought, 'I am taking this one. I earned this one.'"

His second thought, though, was "Katie." Katie Griboski, from the meeting. He assigned this one to her, as she was the most qualified investigator. Punching buttons on his computer, he tries to get it to link up to the Internet. He spends a few minutes fiddling, doing what I'm not sure, like the people at rental-car counters who type for fifteen minutes after asking whether you'd like to buy Hertz's insurance.

"I can turn my phone into a hotspot," I say, picking it up and tilting it sideways, like a dog's quizzical head.

"I love you young people," he says. "You know all the tricks." Finally, the report pops up.

"It concerns the disposition of a biological entity," he says, once again an officer. "*Remains*," he adds, lifting his eyebrows. "So we *do* get good cases. It's not all silliness and lights in the sky."

Since I'm not myself a MUFON-approved researcher, when I ask for a case update a few months later, Wilson says all he can give me is the case number, so I can look up whatever is publicly available online. When I do that, I discover that the public doesn't get to find out the latest on the case, or see its progress. "The 'public view' of an active case is always limited to the basic report so as not to compromise the investigation," Wilson explains. To those outside the MUFON inner circle, the case exists only in the form of its alleged evidence. The collection of alleged evidence could add up to anything. Hypotheses about what's going on up there are only for the exalted.

Still, I look up the case number in the database to see what the original report actually said. It was from a man who'd met a mortician. The mortician had talked about finding an autopsied alien body stashed in a child's casket, beneath a mortuary. Roberts himself had written a thirty-page account and published it with a small press, connecting the body to those allegedly associated with the infamous Roswell UFO crash. The reporter had—after ringing a number of libraries—discovered a copy of this book, and also heard Roberts talk about "visitors" he'd met twice as a child, and a foreign object he'd found embedded in his ankle.

"He has no photos," said the submission, "only memories and a book describing his strange experience."

Toward the end of our conversation, I mention to Wilson that I'm planning a trip to Area 51, and his pupils dilate. "We need to talk," he says, though we already are. By the time we get around to talk-talking about this, though, the coffeeshop is closing, and we step outside. The barista sweeps up behind the door, and a few people congregate next-door, at a punk restaurant/bar called Sputnik—the kind of place where the wood smells like old spilled alcohol, and it's always a little dark, even in the middle of the day. You know, Soviet. I can see the audience paying periodic attention to Wilson's words.

One time, he recalls, Area 51 guards drove over in trucks in the middle of the night. They circled his tent, their lights playing on the fabric like sweeping stage lamps. Another time, they pulled him out the window of his car—or tried. He was too bulky. He egressed through the door, and then they zip-tied and beat him. They put him facedown in the fine dust, so he had to suck it up into his nose, his mouth. When the sheriff came, he threw him in the back of a pickup, whose topper greenhouse-effected him to near heat-exhaustion—his face there, too, pushed into the floor so he wouldn't see what was happening.

Never take your own car out there, Wilson warns. You'll get flats from someone shooting out your tires or from tire-traps *They* set. If you leave the car alone, *They'll* just take it. Never take a rental without getting full insurance. If you're hiring a fixer, watch out for this: Unscrupulous local guides will charge you a couple hundred for a Jeep tour, then strand you there unless you agree to pay five hundred, a thousand more.

"Yikes," I say.

I'm not sure what to believe, though. While I'm very familiar with the idea that people with power—guards, police, the military, your boss—are prone to abuse that power when no one's watching, I also feel like there would be more public outcry if a bunch of middle-aged white guys (the typical Area 51 tourists) actually had their legal rights removed while visiting the desert. Plus, why would people who don't want you around in the first place take your car or pop your tires, leaving you stuck where, by virtue of your stuckness, you might accidentally see something you weren't supposed to—the very things they're trying to keep from you?

These stories might be exaggerations, permutations of the truth, misinterpretations, or misrememberings. Or maybe not. Wilson notes, for instance, that this was in the 1990s, and that the organizations operating in the area now may do business differently.

As soon as I get home, I write to the local guy who agreed to drive *me* around Area 51 in his Jeep. Telling him what I've just heard, I coyly ask if he's heard of any guides stranding people in the desert (hinting in the pauses that he'd never do that to me . . . right?).

II

Griboski was with Wilson on one of his Area 51 expeditions. Yes, she says, the tires went flat and they had to send someone to town for help while she and Wilson survived on Twizzlers in the shade. In a suburban tavern called It's Brothers Bar, she pulls up pictures from the trip, including one of a dragonfly zooming around their Jeep. It *followed* them. Strange as it seems, dragonflies and damselflies do live in the desert, but Griboski believes this one was a surveillance drone, like the insect-spies the CIA (truly) started developing in the 1970s.

We're nearly the only people in the pub, located in a first-ring Denver suburb, home of condos sent from an early-2000s vision of the future and of newly constructed town centers that are actually just outdoor malls.

Next to Griboski in the booth is a backpack with MUFON patches all over it. Many come from the same family as the acronyms on Wilson's card. The most important one is the Special Assignment Team, SAT, a coalition of the best MUFONers from all over, who take on the most promising cases. And, back in the flush times, she was part of a team that Robert Bigelow funded—perhaps with AATIP money. "I miss those years," she says.

She has also missed an appetizer called gator balls, which is this restaurant's specialty. In their center, they contain a pepper jack cheese and jalapeno mixture. That's wrapped in a sphere of chicken, around which are weaved strips of bacon. You can dip them in the peppercorn ranch dressing, which she does not and I do. It's a lot of animals in one object. An ark, of a sort.

"My story kind of started back in the mid 70s," says Griboski, who, despite being old enough to have a grandson, which she does, retains youthful energy and might be cooler than you.

Her own mother was, back in those years, friends with the owners of a ranch in Elbert County, about an hour southeast of Denver. It's a plainsy part of the state, with open, flat spaces. The sky looks so big that it could contain the whole universe; the cumulus clouds pile endlessly against each other.

"Strange things happened out there," Griboski says.

At the time, she wasn't sure what the strange things were, because she was around twelve. At that age, the world remains largely hidden from you, revealing itself in snippets of overheard adult conversation. But whatever form the strangeness took, she knew that police were involved, and that cows were dying in dreadful ways.

After they stopped going out to the ranch, though, Griboski mostly went on with her life. She grew up, she had five kids, she got a degree, she forgot. She was busy, you know? But then, in middle age and after she started watching *Ghost Hunters*, she started to wonder, what *was* all that stuff from her childhood? She began investigating, and she learned these bovine deaths were allegedly part of a spate of "cattle mutilations"—a kind of livestock demise that happens seemingly sans reason—similar to the kind that were inflicted upon the Shermans' cows. Ranchers find cows with organs (especially reproductive ones) removed, often with surgical precision, sometimes blood sucked out, almost always no prints around. The violence generally seems mysterious enough that while some attribute it to normal predation, many others see something more sinister. Cult rituals, government tampering, or—of course—aliens.

Today, such scares have largely fallen from mainstream consciousness, but back in the seventies, they made big news—and even got the FBI up in arms. Of the events near the Elbert County ranch, the *New York Times* wrote, "Who has been killing cattle in Colorado and at least 10 other states the last few months?" in October 1975. Since April of that year, the article claimed, 196 cows (and one buffalo, one horse, and one goat) had shown up mutilated in Colorado. Seventy-three appeared (or, really, disappeared) in Elbert County alone. There were federal inquiries.

None turned up direct indication of human involvement. Colorado investigators concluded that "many of the cattle have died of natural causes. Predators such as coyotes and foxes, rather than humans, then moved in to remove the organs, they believe." But state studies also showed that "a sharp instrument" had sometimes been involved, which, one will note, foxes cannot wield.

Amid these serious investigations and half conclusions, tensions ran high: Ranchers' livelihoods were at stake, authorities had found no readily identifiable cause of all the death, and Coloradans possess a libertarian streak. Livestock groups, among others, offered thousands of dollars in reward for information. Also, they took on potential perpetrators. "Ranchers in Elbert County have reported numerous sightings of helicopters at night," said the *Times*. Fear of being shot at by angry stockmen, the paper reported, led some helicopter operators to delay their flights.

But while the article noted suspicions of government involvement, it also tacitly acknowledged that more than terrestrial theories were flying around. "Asked about possible mutilators from outer space, [Carl Whiteside of the Colorado Bureau of Investigation] replied, 'I'm not ready for the U.F.O. theory. Maybe I'm narrow-minded.'"

Maybe, Griboski would say.

By now, she's bisected her gator ball. Cut open, it leaks fluid onto her small plate. She didn't have explanations, either, but she made it her quest to find some. It was near the beginning of that quest that she discovered MUFON. "I became an investigator right away," she says. She started taking on cases, becoming integral and integrated into the community while keeping the old ranch, and its strange stories, top of mind.

"I love the people and all that good stuff," she says, "but it was really always just about finding answers for what happened." About putting together the pieces whose shape she barely knew.

Soon, she came across *Hunt for the Skinwalker*, which has a chapter about other paranormal hotspots—including one in Colorado. As she read the anonymized account in the book, Griboski felt a connection, felt certain the events were those that had happened on the ranch. Fleets of flying saucers, heavy footsteps where no feet were, humming of the electronic sort. Weirdly dead cows, sketchy Bigfoot sightings, "a nearly seven-foot-tall being in a space suit, complete with cosmonaut helmet," blonde and ethereal strangers in flight clothes. Some suggested it was aliens; some said it was the nearby military staging mind manipulation experiments; some said it was a mass hallucination.

Griboski felt like whatever else it was, it was the place she spent so much time growing up. So she wrote to one of the original investigators: a University of Wyoming psychologist named Leo Sprinkle.

She told him the location of the ranch, and asked if it was the place he'd researched all those years ago.

The innards of Griboski's gator ball have, uneaten, now begun to congeal. She gestures with her fork when she informs me that, yes, the book was talking about the spot in Elbert County. Sprinkle had the original files in the university archives. Soon, Griboski was getting in touch with nonexperts, people who lived in town whom she'd found on Facebook, and who wanted to talk about their decades-old experiences. "It went from 'I don't know if this happened' to 'holy cow,'" she says. ". . . Now the whole town wants to talk, so this is my baby right now."

And she does mean *her* baby. She's doing it outside of MUFON, because once MUFON gets a case, MUFON owns the case. And this one belongs to her. "I've given up everything I do for this," she says, putting her utensils down although she is not much using them anyway. "And I'm like, 'Why?'"

She studies the phenomenon nearly full-time now, having largely stepped away from her graphic-design career. She needs to understand what happened in her childhood, and to piece it together with other people's glimpses and snippets. There is, for her, a very real magnetism to all that. That's the first reason she's into ufology.

"Number two is the people," she says. "It's hard to find real down-to-Earth people."

In looking up to the skies, she's finally spotted them.

"Why are there certain people seeing these all the time?" she wonders aloud. "That's something I've been thinking a lot about lately."

Science-minded skeptics would say that these frequent flyers are primed to interpret whatever they don't immediately understand as alien. Or that they're hallucinating. Delusional. Paranoid. Griboski has other ideas, though. Maybe people manifest their intention to See Something. Maybe they're just open to it. "I've come to believe it's because they expect something to happen," she says, "but if you're a debunker and you don't expect something to happen, you kind of create your reality." (This is a slightly pointed comment, as a few minutes before, she'd asked me if I'd ever seen one, and I confessed that I had not.)

"That's something I hope science can answer," she says.

Science isn't exactly looking into that. But it does occasionally deal in the psychology that interests Griboski. She notes the overlap between people who've had traumatic experiences and those who regularly report paranormal experiences. Scientists might say (have said) a perceived alien abduction is sometimes a displaced memory of human trauma. That alien you remember leaning over your bed is your brain's way of distancing itself from your childhood assault, for instance. If the perpetrator is from another planet, the experience becomes somehow easier to understand than one in which a member of your own species does the same thing. A scientist might also note that trauma survivors have more mental health difficulties, which might sometimes manifest as hallucinations or fabrications.

But that's not how Griboski necessarily sees it. Trauma, she says, sharpens your senses: It makes you more aware, in fight or flight mode, of your surroundings. Your senses pick up on subtleties—subtle UFOs, wispy ghosts—that others, their senses dulled by a placid life, miss.

Her colleagues at MUFON mostly don't buy this. She calls them, now, her "scientific friends." "They're like, 'Why do you think that? How do you come to that conclusion? That's just anecdotal evidence,'" she says, laughing. "I love them. They keep me grounded, because you can go off the rails too."

But she says she's going to prove them wrong someday. Maybe someday soon. "You probably heard this quote from Lue Elizondo," she says. "'A year from now we'll be having a significantly different conversation.'"

It's something he said—without providing any evidence—at 2018's MUFON Symposium, an annual conference that thousands attend.

Sure, he's talking about some kind of capital-D Disclosure. And most of his audience assumes that revelation will be about extraterrestrials. But Griboski thinks bigger. "I think it's about the disclosure of what our reality really is," she says. ". . . Forget the aliens and the craft. It's the nature of reality."

She believes the revelation will connect Skinwalker, Bigelow, Elizondo, AATIP, BAASS. They'll link up into a sense-making circle, and we will all become part of the community.

But who knows, maybe not, she continues. Even if that seemingly sanctioned disclosure doesn't happen, Griboski might stumble on something good on her own. After all, she's given up everything for this.

"Here's what I'm hoping," she says. "I'm hoping I get privy to some really good information, and they pay me off to shut me up."

She imitates herself then: *UFOs don't exist. There's nothing to it.*

If I see her in a year and that's what she's saying, she says, I'll know.

III

There is a gap between what some MUFON members believe and how they characterize their investigations. Griboski believes I don't see UFOs because I don't want to—which implies they'd be there for me to see if I did.

Like anyone, she's capable of setting aside those personal beliefs when she picks up a case file. After all, scientists who believe God created the universe can aim a telescope back toward the Big Bang and draw conclusions that have nothing to do with a long-haired nice guy. But when any human is involved in any endeavor, they cannot help but influence it. Scientific methods are civilization's so-far best attempt at removing biases, but nothing that involves a person (and probably nothing that involves a robot) is ever truly objective.

Still, James Carrion, a former international director of MUFON, sees more human influence on the organization's UFO research than he'd like. Carrion discovered MUFON in the late nineties, when he moved to Colorado. MUFON was headquartered there back then, and Carrion would drive down to volunteer. He works in IT, so when he saw that MUFON stored its old case reports in cabinets, it seemed like a missed opportunity. Why not digitize them, make them searchable and available to more people? He proposed what came to be called the Pandora Project—a scanning plan that would release the whole history of sightings from 1969 onward out into the wider world.

After the success of that effort, Carrion became a board member, and when director John Schuessler—who'd helped Andrus form the organization many years prior—retired, Carrion threw his name into the candidate lineup.

During his time at the top, and deep in those old investigations, Carrion saw things that gave him pause. "There were a lot of cases that were in the files that were labeled 'unidentified,'" he says. And while some seemed genuine mysteries, others seemed low-quality. "I was a little taken aback by the lack of uniform investigation," Carrion continues—and taken aback by how often belief got in the way of evidence.

As an example, Carrion cites some filming MUFON did for a television show called *UFOs over Earth*. The crew traveled up to Pennsylvania to visit a woman who said a UFO had zipped over her house, dropped physical evidence on the leaves of her tree. That would have been great—if it were true. But an investigator's job should be to essentially think of a sighting as guilty till proven innocent, explainable till proven duly weird. "But the field investigators were way too eager to believe her story," says Carrion. They took samples from the trees and sent them to a MUFON-affiliated scientist whose confirmation bias, Carrion says, distorted his analysis. "Their beliefs come first," he adds, "and the science comes second."

It went all the way to the top. "Even the MUFON board of directors showed their beliefs to be opposite to what true scientific investigation is," says Carrion. From the base of the hierarchy to its peak, MUFON members fell before confirmation bias. "If something supports the belief, then great," he continues. "You're going to go push that story and try to promote it. If not you're going to discard it."

When Carrion was at the top-top, he tried to change that, to beef up and homogenize MUFON's methods and standards. "If MUFON is going to be taken seriously, we need to take investigation seriously," he recalls thinking. "I basically rewrote the field investigator manual to be more evidence-oriented." Would-be researchers have to study this manual and pass a test on its contents, and then follow its precepts and procedures when they go out in search of answers. It's hard to tell how much of an effect that had on the quality of analysis, but Carrion also instituted a yearly training at the annual symposium, where they could all learn the same standards.

After that 2008 standardization, though, Carrion didn't stick around long to see whether the everyday MUFON leader—or those atop the pyramid—stuck with more rigorous work, or whether rigor made a difference in the number of actually unidentified objects.

Carrion, in the end, left not over standards of evidence or soundness of conclusions but over politics and Robert Bigelow. When Bigelow approached MUFON in 2008 with the idea to form an elite, supercharged, fast-response team of investigators—the STAR team, perhaps part of the Pentagon's AATIP program—he didn't approach Carrion, who was director at the time. He approached Schuessler. "[Bigelow] said, 'I have this sponsor who is willing to put in a lot of money to basically rent MUFON,'" says Carrion. Bigelow, and The Sponsor, wanted to benefit from the scientific data that came out of the investigations. "'We want to find out what the technology is behind UFOs, because we want to achieve some breakthroughs based on terrestrial approaches,'" Carrion summarizes.

He wasn't happy that Bigelow approached Schuessler rather than himself. He also wasn't happy that anonymous backers funded the proposed arrangement. He wasn't happy that Schuessler was the only one who knew the identity of the backers. But MUFON took the deal, which basically paid the organization a monthly retainer: When a "high-value" case came up, high-value investigators like Griboski could get on planes and go check them out. "We had more money coming in than was going out, because we didn't have that many high-value cases," claims Carrion. "Money accumulated in the bank. Bigelow, halfway through, said, 'We don't like that.'"

Carrion didn't like that Bigelow didn't like it. But Bigelow, used to going over by flying around, convinced influential board members to rewrite the contract.

Carrion's frustrations and doubts compounded. He had heard stories about Skinwalker Ranch, and about the people who worked there. They were perpetuating old-school UFO propaganda, like stories involving Bob Lazar—a person who claimed to have worked in Area 51, reverse-engineering a spaceship. Even within ufology, most people didn't and don't believe Lazar's story, which is as porous as limestone. (To the Stars, Inc., published Lazar's autobiography last October.) When Carrion went down to Skinwalker to learn more, Bigelow refused him entry. "I started thinking that something is not right here," says Carrion. "If Bigelow is who he claims he is, and he's doing this for the reasons he says he is, he shouldn't be talking about Bob Lazar, asking us to sign nondisclosure agreements, denying us access, keeping from us who his benefactor was."

Carrion resigned in 2010, putting out a public statement—titled "Strange Bedfellows"—on his blog. In it, he revealed his qualms about the Bigelow deal and hinted that the government was involved. Who was the true sponsor of the STAR team? "It is time for MUFON to sweep its own house clean," the letter concluded. The cobwebs wouldn't get clear for years—not even when Robert Bigelow landed on the front page of the *New York Times*.

Carrion learned about AATIP at the same time as everybody else, when AATIP—in which he had perhaps been a somewhat unwitting participant—became an international news story.

Today, he doesn't do ufology anymore. "I've closed that chapter of my life," he says.

Immediately after his resignation, he pursued three historical cases that interested him and wrote two books—*Anachronism* and *The Roswell Deception*.

"Okay, I'm done," he recalls thinking. "Enough of my life wasted. Time to pursue more meaningful endeavors."

And just like that, James Carrion was gone.

IV

Most of the people I have spoken to for this book are the independent researchers who toil away among gigabytes and peta-piles of documents, who go out searching—for historical context, for present conspiracies, for future technology—on their own. But for those who want to find their people, MUFON fills two very human voids: It provides both an accepting community and an earnest search for answers.

Neither of those yearnings is new or unique. The first is why we have, well, civilization. The second is why we have religion, philosophy, science, literature, art, late-night conversation. You might not agree that we need answers about *UFOs*, but it's hard to deny the very human need to understand the world around us and the universe around it. Humans have always sought to explain what seems inexplicable. In the distant past, people tended to attribute strange phenomena—earthquakes, eclipses, fireballs—to angry gods who demanded appeasement. Today, some people still invoke a deity or paranormal power of a different sort to explain such terrestrial machinations. Others invoke science.

Some straddle both worlds. And that seems to be where ufology resides—or, rather, where it envisions itself residing. Likely, neither religious people nor scientists agree. Religion mostly requires Earth and humans to stay central. Throw a bunch of smart aliens into the mix, and you have to ask who their gods are, who created them, and whether they will be saved. Is our Jesus theirs? How did he travel so far so fast? Clergy don't typically like that line of inquiry.

With the scientific establishment, meanwhile, ufology has a twisted relationship. Ufologists sometimes distance themselves from scientists, disdaining scientists' disdain for their work, proclaiming scientists' methods and standards to be narrow-minded. At the same time, they want to imitate the scientists. As political theorist Jodi Dean wrote in *Aliens in America,* summarizing the thrust of a book by ufologists, "By becoming like 'real' science, ufology will attract it and, in effect, merge with it."

For that to happen, though, science as a cultural institution would have to change and accept explanations and evidence according to different standards. Till that happens, science doesn't know what it's talking about, and MUFON does.

At the same time that researchers shun the institution's perceived biases, though, they seek its approval and revere its authorities: When someone with a PhD or a study in a peer-reviewed journal nods at UFO research or extraterrestrial visitation, many ufologists say, essentially, "See?" When a symposium speaker has worked for NASA or had a metal sample tested by a university professor, they say, "So you know it's legit."

Still, lots of skeptics point out that MUFON symposia and local meetings regularly feature pseudoscientific content (alien-human hybrids, the Secret Space Program, extraterrestrial abduction). Those featured topics hint that when MUFON says "it's not aliens, but" the emphasis may be on the "but."

If you spoke to scientists, many would likely say the problem is that the organization takes witnesses at their word. There's very rarely physical evidence; there's no way to replicate the "experiment"; there's just the account of an imperfect person. In reality, no human (including those scientists) is a reliable narrator. Memory and perception are decidedly faulty. Even the best human brain is a great distorter, an A+ inventor, an impressively illogical filler-in of gaps. Seminal and controversial work by psychologist Elizabeth Loftus, for instance, has suggested that telling people they have a memory of an event can *give* them that memory—even if the event never occurred. "People can be led to believe that entire events happened to them after suggestions to that effect," Loftus wrote in a 1995 *Psychiatric Annals* paper. If you can create a false memory for yourself out of the thin air of your brain—a recollection of being lost in a mall as a kid, in Loftus's subjects' case—and truly believe it to be true, it's not hard to imagine that you could misremember details of something that actually happened. And other studies, in fact, prove this out. Even so-called "flashbulb memories," those hard-imprinted recollections of where you were and what you were doing when you heard some shocking news, can go wrong—when many had thought that the emotional resonance of the moment set the memory in gray stone. Take a 1992 study about students' flashbulb memories of the *Challenger* shuttle explosion: When researchers compared their recollections right afterward and then around three years later, "40% of the informants were clearly inconsistent across the two occasions." Similar mix-ups may be true of eyewitness testimony in court: The Innocence Project, which seeks to exonerate the wrongfully convicted, estimates that "75% of wrongful convictions that were overturned by DNA testing involved erroneous identifications from victims or witnesses."

Memory mishaps present a problem for UFO sightings—as do straight-up inaccurate perception and interpretation. For instance, in unfamiliar situations, the brain seems chemically and functionally primed to say "Agency!" when it sees ambiguity. If we see a menacing shadow behind a tree, it's safer to imagine it's that of a wolf than that of a rock. If we're wrong about the wolf, no big deal. If we're wrong about the rock, we're dead. Today, most of us don't encounter many wolves. But if we see a strange light in the sky, our brains tend to interpret its movements as being under intelligent control. Guided by *something* with intent. It's similar to the effect that—on an organizational level—leads people to infer conspiracy. *Something, somebodies* more powerful than us must be in charge.

If that doesn't lend enough doubt, combine it all with the fallibility of our memory. Eyewitnesses standing side by side at a crime tell different stories. When we recall a memory, our brains may rewrite some of the details, updating what we thought was fixed, in a still-uncertain process scientists call "reconsolidation." It's perhaps meant to make us and the files in our brains most adapted to our present circumstances, but it can also just make us wrong about the past. We tend to collapse time and mix together different events (did we get Funfetti cupcakes on the trip to Phoenix in 1999, or the one to Pittsburgh in 2001?), misattribute remembered conversations to the wrong chat-partners. The raw information that feeds into our sensory organs has to be processed by our brains' algorithms, which were forged by our unique social, economic, geographic, political, historical, cultural circumstances. It's a wonder we trust anything that comes out of anyone's mouth, including our own. And, you know, maybe we don't. Maybe there is not a consensus version of modern reality, just a herd of individual perceptions. "'Common' sense is lacking," Dean wrote. "There are only particular senses."

And peculiar senses.

Attend a MUFON meeting as someone with a genuine UFO interest, though, and you're likely to find you have particular, peculiar senses in common with the rest of the audience. The world coheres.

April 2019's Colorado MUFON gathering takes place at a high school actually called Colorado's Finest High School of Choice, which is much emptier of volleyball players than the last one. When Wilson again steps to the front and asks how many people have never attended a meeting before, a fair number of hands shoot up.

"We hope this is something where you go, 'Wow, these people are not the idiots I thought they were,'" he says.

He then invites them to the upcoming annual symposium, where "nobody thinks you're a crackpot," and to join MUFON, because "you have an opportunity to have friends everywhere."

Soon, his welcome over, Wilson cedes the floor to today's guest speaker, Major David Toon. He's a retired army guy, his talk titled "The Search for ET Vehicles in Solar Satellite Imagery."

Hiding out near the sun is a terrible idea for a spacecraft. It's hot there. It's irradiated. Things melt and fry and malfunction. But no one, including me, says that. We smile and wait to hear him out.

"This feels like home," Toon says, before he even begins.

SEVEN

RIDING THE EXTRATERRESTRIAL HYPOTHESIS HIGHWAY INTO AREA 51

I

'm thinking of Wilson's warnings when the rental Subaru—our "tent" for three cold nights of camping at Area 51—pulls into the town of Caliente, Nevada. Because as soon as the Subaru pulls into the town of Caliente, Nevada, we see a great unknown *something*. Exactly the kind of something, in fact, that we'd come here to see.

"Look," my sister Bekah says from the front passenger seat. She leans toward the window and cocks her head. Following the vector of her tilt, I see it too: a group of four lights—orangish, reddish—clustered into an oval, a saucer keeping still above the small city. But as soon as my eyes have passed the information to my brain, these lights—and the shape they circumscribe—fade out. Gone.

"Holy shit," I say.

"Holy shit," echo my sister and the other passenger, satellite industry analyst and friend Carolyn Belle.

Bekah's next words are "It's the aliens!" although none of us actually thinks it is. Still, it stuns that as soon as we neared the border of the much-hyped Area, we saw precisely the sort of object that people have been claiming to see for decades. And although it perhaps shouldn't have been a surprise, it was: In my doubt about other people's senses and dedication to truth, I had assumed we wouldn't see anything at all, or that anything we *did* witness would not be so strange as to confuse. But here I was, talking about a disc in the sky that just *disappeared*. Here I was, a person who could now say, "I saw something!"

As we leave this last outpost before the high-tech wilderness, I recount for my passengers what Doug Wilson, of MUFON, warned me about: That the Area 51 guards might harass us, maybe violently. That someone will shoot out our tires as we drive, or push our heads dirt-down. That our guide—a local named Joerg Arnu, who runs an informational Area 51 website called Dreamland Resort—might drive us out to the middle of the desert and demand triple-digit cash to return us to civilization.

In my retelling, I brush this all off, pointing out that unless we break the actual law, we have nothing to worry about (the passengers, perhaps having had less time to digest the idea of an assault by authorities, are not so confident). But I continue with what is for sure true: that the military forces inside the advanced aircraft testing facility are, truly, watching what curious tourists do on the roadside pull-offs and just-over-the-line peep spots. That sensors and cameras aim themselves not just at the latest, greatest jets but also at visitors like us. That if we do cross the site's perimeter, security could technically kill us without question and face no legal consequences. Most importantly, I know that They (the aggregate, occluded plural) do not want anyone from the outside to know what happens inside Area 51's boundaries. It's a wish that will get granted. We will never (no one outside will ever) even get close to finding out what's going on there, at least not until today's secrets are declassified fifty years from now.

As we continue to drive down the long, straight road to this strange base that didn't officially exist until 2013, more orange lights appear. They always hover for a few seconds over the horizon, and then flicker out. We shout and point each time, like we are playing the speculative-technology-themed version of Car Bingo.

Needing an explanation, I make up a hypothesis: These lights are aircraft dropping in and out of stealth mode—turning their lights on, then going dark as the night itself. Maybe they are playing an aerial game of paintball, I reason. A bunch of jets chase one another, and when one *finds* another, it shines a *gotcha* signal, and then they both go back to hiding and seeking in the open skies. My peers have no better ideas, so like uncomprehending humans throughout the ages—those who thought volcanoes spit fire into the air because the gods were angry, those who thought disease spread because of "bad air," or miasma—we accept this inadequate one.

Soon, the headlights reflect off a green road sign: EXTRATER-RESTRIAL HIGHWAY, it reads, in a font anthropologist Susan Lepselter calls "vaguely 1980s 'computer-modern.'" This sign, and the state-official designation backing it, appeared just in time for the movie *Independence Day*'s debut in 1996, notes Lepselter. She spent time living and waitressing here as part of research for an ethnography called *Resonance of Unseen Things*. "While the federal government was trying to keep the public's eyes away from Area 51," she said, "the state of Nevada was stoking the commodifiable desire bred by high-tech military secrecy." In military areas like this Area, an inherent tension exists between the black-budget parts of the government and the almost necessarily poor and rural regions they operate in, which may wish to capitalize on their infamy. But like a true American raised in a late capitalist society, I pose for a picture in front of the physical manifestation of this tension, spreading my fingers into the peace sign for the camera.

When we crest Hancock Summit, the last high point on the way to our destination, we look down on the darkened Amargosa Valley. This plain has menacing topography. It lies a mountain ridge over from Groom Lake, where Area 51 is. Also, it's near Yucca Mountain, where since the 1980s the government has on-and-off planned to deposit the nation's nuclear waste, making it potentially uninhabitable for tens of thousands of years—so long that the Department of Energy had to hire linguists and anthropologists to determine how best to communicate "Skin-melting poison! Steer clear!" to people who won't be born till 12,020 c.e.

Waste from human violence and power make a good metaphor, but you can't actually see Yucca Mountain from here. And given that it's nighttime, we can't see any other mountains either. The geography here, though, looks strange to the eyes of people who grew up "back East." There, mountains are gentle and grow leaves. Here, they are sharp mounds of semi-permanent dirt, hydrodynamically eroded into caverns and cones, rising from nothing. Their low brush, cacti, and Mormon Tea plants feel exotic, their beauty austere and distanced. Over there, mountains seem to *want* life to exist. Here, everything about them (even without the gun-toting guards) screams, "Get out!" And, because "hard to get" has, historically, held a certain appeal, I like them a lot.

But all these contours pass us by unseen as we search for roadside pull offs where we could camp. Much of the property around Area 51 belongs to the Bureau of Land Management (BLM), the governmental organization that deals with one-tenth of the country's landmass, some 245 million acres. On that acreage, you're generally allowed to camp anywhere you're not contaminating a water source, of which there are not many here.

Finally, we choose a pull-off that Arnu has labeled—on the map he hosts on Dreamland Resort—"Gravel Parking." It's a wide rectangle with rounded corners. An interpretive sign on one side explains that there are two kinds of Joshua trees: tall and lanky ones, short and squatty ones. Just a single species of moth pollinates each. Without this hyperspecialized symbiosis, these trees wouldn't exist at all. What relationships of convenience might alien environments have forced on other planets? Just as I'm contemplating that, head-lights from a distant car appear down the road. They are the first ground-based illumination we've seen in a long time.

"Let's wait for it to go by," says Carolyn, wary of setting up camp when a stranger could whiz by and note exactly where three women are camping alone.

So we get out and kick around some pebbles, pretending we are about to head somewhere else in the universe. But the lights seem to stay exactly where they are. You could say they *hover*, if you were inclined to use such a word. Then, with nary a flicker, they disappear.

Maybe the person has turned off onto a side road, or perhaps to their own gravel campsite.

"Turn on your lights and pretend to leave," Carolyn urges, craning toward the window. "To see if they turn theirs back on."

I do, pulling the car back toward the road and putting on my blinker as if such an indicator of intention matters out here. And as I do so, the lights return, just as bright, at the same seeming distance, with the same eerie stillness.

"What the *fuck*?" we all say.

When we pull back into our place, the lights disappear again.

"Maybe it's something with the guards," I say, recalling Wilson's words. I believe his accounts came from *somewhere*, from some starting point that perhaps mushroom-clouded into more. It doesn't seem crazy to think that the road has sensors, and that if you trip one, *They* might let you know you have company.

Then the lights reappear, immobilizing us. Their action-reaction nature makes it seem like the car—or *Whatever*—knows we are here. And whether it is a random pickup or guards or a tripwire-activated light, that feels unnerving. But underneath that—almost so subliminal that you'd need a therapist to unearth and articulate it—another idea lays buried: Whoever they are, they care enough to notice us. We're special. That's how people who see UFOs often feel. As anthropologist Benson Saler put it, "The others, who are presently alien, are not indifferent to us."

But soon, we hear an engine, and in a minute or so, an eighteen-wheeler whooshes by and is gone, taking the lights with it. The air cartwheels around us, and darkness returns.

Laughing off our folly, we settle into our temporary home on the Extraterrestrial Highway, building a fire to ward off the creeping cold. We shout triumphantly at each new craft that shoots through the sky. Our vehicular revelation—that we had imbued something benign with false agency and garnished it with threat—has stripped the desert's sky-based strangeness of menace.

Orange orbs continue to fizz on and off. Huge fleets of white blinking lights come up out of nowhere and take over a whole hemisphere of sky. Rumblings come from the edge of the earth and register in the sternum.

We are not afraid: We are elated. We whoop and yell at the sky. A very small, very arrogant, part of me wants to believe, though I know it isn't true, that this too-perfect celestial show is for us.

II

In the morning, we look at what the night had kept unseen: just a regular desert. Desolate and magnificent, sure. Stunning, yes. But not much different from a thousand other spots in Nevada, Arizona, New Mexico, Texas.

What I mean is that in the light, this area does not seem special. And neither do we.

The highway stretches long and straight for miles and miles and miles. But it does have a dip and a curve: the two places where the truck's lights had disappeared. What dupes we were, I think, how prone to read meaning into coincidence, even when we knew better.

That morning, we continue along Highway 375 toward the town of Rachel, traveling 20 miles before the scenery changes at all. But finally, we sweep over another pass and reach the "Welcome to Rachel" sign, which is written in what appears to be Comic Sans. Friendly. Approachable and unserious and maybe a little backward. A matte-black saucer flies over the "e" in "Rachel." "Population," the sign continues, "Humans: YES. Aliens: ?"

This would seem strange if you accidentally passed through, knowing nothing of Area 51's proximity. After all, when you first cross into the town limits, it looks a lot like any other small, desert spot: dusty one-story houses and trailers, battered by the wind, blasted by sand, dimmed by the ever-staring sun. But then, you see something slightly different: a sign advertising a restaurant-bar-motel, above a picture of a *Close-Encounters*-style alien.

"Earthlings welcome," it says. "Little A'Le'Inn." (Get it?)

Next to the sign, a dull metal saucer swings from a crane. Trailers cluster around like a herd of lesser intelligences. These trailers actually *are* the motel. (Before the trip, when I had asked Arnu about places to stay, he called this one—the only one, mind you—"rustic.")

Inside the bar-restaurant-souvenir-shop part, customers can buy heather-gray and army-green T-shirts that say "Restricted Area: Deadly Force Authorized" and "I had a drink with an alien." Shelves host shot glasses etched with saucers. "Alien Vodka" holds a place of prestige in a corner.

Despite its current appearance, this place wasn't always an alien shop. And Area 51 wasn't always an alien place—though it has been a *strange* place since the early 1950s. Back then, it hosted the Nevada Proving Ground, now the Nevada Test site, where the Atomic Energy Commission tested nukes, blasting holes into the ground and mushroom clouds into the sky. Even citizens of Las Vegas could see the wide clouds and feel the ground shake. The people who lived nearby, even those close enough to feel the deleterious effects of fallout, were told almost nothing. The area was, in other words, already plenty shrouded, plenty protected, and plenty tied up with military-industrial interests before Area 51 came along in 1955.

People have known for decades that something was going down out here in the desert, and that their government wasn't telling them the whole truth. And that was true. But that something didn't usually include aliens. One man changed all that—a man named Bob Lazar.

Initially calling himself "Dennis," Lazar did a 1989 interview with none other than George Knapp, of *Hunt for the Skinwalker*. In the segment, he told an unbelievable story. He'd worked right next to Area 51, he said—which people knew existed but which the government had not and would not for years yet acknowledge. There, Lazar claimed that he reverse-engineered extraterrestrial flying saucers for human use. These saucers were fueled by a stable version of element 115. In a series of interviews, during which Lazar eventually revealed his true identity, he also claimed that aliens from Zeta Reticuli, a binary star system around thirty-nine light-years away, had been stopping by Earth for some 10,000 years.

Given that that's an extraordinary claim, it requires at least ordinary evidence. So since Lazar came out, his credentials and assertions have rightly been scrutinized. And some haven't really held up. He claimed he attended MIT, for instance, but the school has no record of him. Either exploiting unfalsifiability or telling the truth, Lazar says the government tampered with his records to discredit him. And he *did* perhaps work at Los Alamos National Lab, like he claimed. His name appears in an old internal phone book, although the facility denies employing him.

Still, there's no real evidence to support the rest of his story—the parts that don't involve education and employment but, instead, aliens. It's not the kind of story that lends itself to contrary evidence: Have *you* ever tried to demonstrate that there are *zero* flying saucers in a test range that was "born classified," meaning secret since the moment of its conception? Rhetorically and scientifically, it's a losing battle. And we'll probably never know Bob Lazar's truth. Many ufologists consider him a scam artist who got over his skis while telling a good yarn. But for now, the main thing Lazar has dragged out of Area 51 is wider-spread knowledge of its general existence, and the near-universal and dubious association of its cutting-edge technology with the extraterrestrial.

Before and immediately after Lazar gave his accounts, the Little A'Le'Inn was simply the Rachel Bar and Grill. But business being tough, every town needing its Thing, and this being the only town, owner Joe Travis and his wife Pat rechristened the restaurant, just as Nevada rechristened Highway 375.

For a while, the inn hosted a UFO conference, and at night attendees would trek out to sky watch. Arnu and his friend Bill Whiffen would go out in advance, armed with helium balloons and rave-style glowsticks. They'd tie them to the balloons and let them fly before it got dark, so when the conference-goers looked up, they'd see floating neon colors.

The next day at the inn, the attendees would always rave about the crazy lights in the sky.

"We created our own little UFO story," Arnu told me later.

Arnu arrives at the A'Le'Inn in a big SUV, pulling up and saying *hi* to the hungover twentysomethings rocking in rocking chairs out front before he greets us.

"You ready?" he asks, and we pack into his Tahoe and head right back out on the Extraterrestrial Highway.

Arnu has owned property in Rachel since the early 2000s. Back in its boom, when the tungsten mine near Tempiute Mountain was still digging wealth out of the planet, around 500 people lived here. Today, it's a small town—just around fifty residents, who meet up at the collective mailbox when the Postal Service arrives. Young people, Arnu says, tend to leave. There's no TV reception. There's just a squeak of cell phone service. Few places exist to build a career, none to go to college. Some people work at what they simply call "the test site," an umbrella term that could refer to any of the secret-squirrel operations nearby—the Nevada National Security Site, the Tonopah Test Range, or Area 51.

Around ten people also work at the A'Le'Inn, by far Rachel's biggest employer. They're always hiring, because people are always leaving. But people are always showing up, too. "Sometimes they come up here because they are interested in Area 51," says Arnu, "and they just get stuck."

That's what happened to Arnu, decades ago now. It all started with online research into Area 51, reading a website run by a former programmer and airline worker named Glenn Campbell. In the 1990s, Campbell ran the Area 51 Research Center and two UFO newsletters—*The Groom Lake Desert Rat* and the just plain *Desert Rat*. The newsletter logo featured a sentient rodent with safari shirt, walkie-talkie, and binoculars, underneath the tagline "The Naked Truth from Open Sources."

Recalling this, Arnu speeds along the straight road. "He was one of the first that brought the attention of the general public," he says. But Campbell was mysterious, evasive. "I wanted to know what's really going on here. Are there UFOs are there no UFOs?"

So Arnu took a day trip, traveling from his home in San Francisco. And when he arrived, he found a place that was fascinating as much for its terrestrial qualities as its celestial hypotheticals. "I had never really experienced the desert in this way," he says. "And it was just like, 'Oh my God, this is a whole different world.'"

He thought of it, thinks of it now, in terms of motorcycle trips—a hobby of his that he just calls "riding." "It's always my thing: I want to see what's behind the next turn, the next hill," he says. And despite how *this* highway feels—unchanging, flat, forever—if you veer from it, turns and hills and the secrets behind them abound.

Arnu went back home knowing he would return. The presence of the place loomed over him, shook him. Soon enough, the labor market gave him a chance: His company downsized, so he took a severance package and car-camped around Rachel.

Soon after that, Arnu started his own website, mostly a blog detailing his daily exploits: As he summarizes it, *Today I went out to this gate, this is what I found, check out my pictures.* More important than anything he wrote, though, were the comments sections.

"It's like people were only waiting for a place to congregate," he says. He soon started a forum—still going strong today—dedicated to such interaction. "We're geeks," he says. "We're loners. But at the same time we also want to discuss what we do with like-minded people."

He moved to Vegas in 2002 and then bought the property in Rachel, working remotely a lot so he could spend a week at a time in the remote desert.

"And here I am," he says. "Years later. Still unraveling the mystery of Area 51."

Arnu looks through the Tahoe's windshield and points at a prominent peak ahead of us. If you can get to the top, you can see inside Area 51, which would then be 26 miles away. This high spot is the only one left with that view, the military having gobbled up all closer vantage points in a series of land grabs. Here's what the base looks like from up there: Dark, if you're doing it right, because the interesting stuff happens at night. But all of a sudden, way across the valley, a runway illuminates itself, a long line of lights dotting the landscape. "You know something is about to happen," Arnu says. Aircraft bulbs streak along the runway, as a *Whatever* speeds to takeoff. And as soon as the *Whatever* is airborne, its lights blink out of existence, and so do the runway's. The Earth becomes as optically opaque as it was before.

It's not that they appear. It's that they disappear.

Nevertheless, the base continues to give away information invisibly: Pilots talk on radios, and if the chatter is not so secret, you may be able to catch a monologue.

Arnu has a radio scanner, which he now turns on, mounted to the dash of his Tahoe. It runs through many Hertz in search of such communication. As the display rolls across frequencies, I prepare to tell Arnu about what we saw last night, feeling silly and like every other overexcitable person who's ever visited the region.

I know from our prior emails that Arnu doesn't ride the alien train. Sure, creepy stuff happens here. Sure, there are strange lights, technologies we can barely fathom. But they don't require invocation of the extraterrestrial: They're just the government, doing things the world isn't privy to—the growing up of projects perhaps born classified, just like it always has here.

That started with the U-2, which flew twice as high as a commercial jet, and much higher than anything else at the time. Workers commuted daily on passenger jets—a secret service people call, in its modern incarnation, "Janet airlines"—partly so that permanent residences would not reveal the scale of efforts here. U-2 pilots, though they worked for the CIA, wore civilian clothes and pretended to do weather-related research, according to the book *Area 51* by investigative journalist Annie Jacobsen.

Later, Area 51 hosted the Oxcart spy plane project, the U-2 successor that also flew close to the sun but showed up dimmer on radar. Jacobsen writes that FAA and NORAD employees were instructed "not to ask questions about anything flying over 40,000 feet." And when commercial flights crossed paths with an Oxcart, and a pilot *did* report it, the FBI would meet the plane at the gate, asking passengers to sign nondisclosure agreements.

Around the country, people nonetheless spotted spy planes and reported them as UFOs. Says a CIA report from 1997, "Over half of all UFO reports from the late 1950s through the 1960s were accounted for by manned reconnaissance flights (namely the U-2) over the United States." Many, including UFO skeptics, dispute this take, but it doesn't seem absurd that the government would use UFO reports to understand how conspicuous its technology would look in less friendly skies. And it doesn't actually want people to see skylights and think "spy planes." So it is sometimes in the feds' best interest to let people attribute the phenomenon to something mysterious, unearthly, *not them*. And—bonus—because many people thought UFOs were woo-woo and not "real," whoever heard about *these* UFO sightings would likely dismiss the very real U-2 or A-12 their kid had just seen. The government's secrets could stay secret. If you wanted to create a theory about why the military hasn't come out swinging against some of its pilots' more modern sightings, you might consider this part of the past.

"'Oh, well, these people just saw another UFO,'" mimics Arnu. "In actuality they may have seen something super-secret . . . If you make people look like fools when they say they saw something, if they say they saw something super secret, what better way to discredit them?" Given the government's history of passive deception, and active secret-keeping, here, is it any surprise that people suspect it could be hiding something *more* inside Area 51?

But I want to know what Arnu, who sees this stuff every day, thinks of my sighting. So I describe the on-off lights, their hovering, and my theory that this was some kind of hide-and-seek exercise.

Arnu frowns in concentration. "Were the lights kind of orange?" he asks. "A bright orange color?"

"*Yes!*" says Carolyn from the backseat. Arnu nods and then goes on to describe exactly what we saw, detail for detail, as if he were there.

"That was flares you were seeing," he says. A plane chases another plane, and the chaser sends off a (fake) heat-seeking missile. The chased plane drops flares, which burn so hot that they distract the missile, which then chases them instead of the jet's exhaust. These planes drop flares in patterns—disc shapes, sometimes—to send the missiles clear off course.

Hearing this incident repeated back, with more meaning, makes me feel the way people do when they discover their seemingly singular experience is, in fact, universal: equal parts relieved and disappointed.

Arnu's first UFO sighting, turns out, was also flares. He had been camping right where we did, in the gravel parking area. "I looked over Tikaboo," he says, referring to one of the peaks, "and all of a sudden, I see this disc-shaped object of orange orbs hanging in the sky."

It's all true, he recalls thinking. *They're coming to get me.*

But they weren't and they didn't. He was just primed: He thought he had witnessed a UFO because that's what he *expected* to witness. "Your eyes see what you want them to see," he says.

He then begins to talk about YouTube videos of cars disappearing on the Extraterrestrial Highway. They're not disappearing, he says: They're coming down from summits, hitting dips.

"We saw that!" I say, and describe how I scared ourselves into thinking that the guards had set a trap.

"That's why I'm such a skeptic," says Arnu. "Because I've seen it. And I know for a fact what they're describing is very explainable."

Talking to Arnu feels like seeing a therapist who understands, even when you don't, that your problems are all because of your mom.

III

Arnu soon pulls over at a spot labeled "the Black Mailbox" on his map. Long a landmark for Area 51 tourists, it was once just a lower-case letter-collector sitting at an intersection. But because there were no visible houses, and because people knew the entrance to Area 51 was down this secondary road, they believed it to be The Mailbox for the secret site.

People began to camp next to it, to steal letters addressed to the rancher—a guy named Steve—who owns a lot of acreage and grazes his cattle on BLM land (locals call this valley "Steve's Valley"). In the late 1990s, Steve replaced the regular mailbox with a giant steel one, padlocked. And though he was annoyed and did want his letters, the tourists also amused him, so he installed a small mailbox with a drop slot addressed to "Aliens." Arnu and Steve and other townies would sometimes come sit out here, slug beer, and read the missives aloud.

Today, the steel box remains, a white board with the words "DROP BOX" painted on it. Someone has arranged a circular shrine of rocks below. As we stand in the spotlight-bright sun next to this shrine, Arnu points over at a peak: Bald Mountain, on which sit one building and two radar domes. These, he says, track test flights out of Area 51 and other nearby installations. A remote-controlled camera up there has eyes on most of this valley, he claims. Later I see a telephoto picture of this camera, lens staring down lens. But then Arnu turns more speculative. "There is a rumor—" he says, "and these exist, I know they exist—laser microphones that you shoot at a surface that picks up sound by vibration of the glass."

He's right. You can use the laser kind of like a radar, where its reflection off the surface of an object reveals that object's vibration, caused by sound. In 2014, scientists from MIT, Microsoft, and Adobe also debuted a "visual microphone" that could reconstruct sound (in the demo, "Mary Had a Little Lamb") from the vibrations the sound waves induced in flexible objects nearby (in the demo, an empty chip bag). A high-speed camera and some trick-pony algorithms are all you need. "The rumor is you're driving by on a highway and they can listen to you," he says.

Maybe they are; maybe they aren't. But the point is that they *could* be. It's not hard to believe, when it seems like we're watched and data-scraped all the time anyway, even far from classified bases.

We get back in the car, bumping down the side road toward the back gate of Area 51. Jet noise rumbles across the plain. We swivel our heads around like *they're* remote-controlled cameras, trying to lock on.

They fly low, Arnu says—like low-low. You have to look right at the mountains to see them. And sure enough, a jet passes right across the ridge in front of us.

"Oh, he's coming our way, actually," says Arnu.

Like one of those *Star Wars* vehicles, the craft veers through a tiny gap in the mountains, its wings looking like they could clip the cliffs. It whooshes over, and then it is gone.

"We may have just been looking at something super-secret," says Arnu, "and you wouldn't even know."

Though it wasn't a secret *plane*, maybe they were testing a new sensor or piece of equipment that the world doesn't know exists.

Arnu smiles and moves his hand toward the scanner. "Let's play Area 51 music," he says.

The machine hooks into a pilot broadcast, probably from the flier we just saw. Arnu looks visibly pleased, having made a connection with the area's unseen secrets. Then, just as soon, the pilot's broadcast is gone.

After a few minutes of scanner silence, though, BEEP comes over the speakers.

"Road sensor!" Arnu says gleefully.

The facility has placed these at intervals from here to the gate, alerting authorities to passersby who could become trespassers. Arnu and a friend named Chuck Clark first discovered them around 2003, after realizing they emitted the radio pings we just heard.

After the two friends learned how to locate the sensors, they did so purposefully: mapping them, digging them up, taking them apart, snapping photos, reassembling them, replacing them.

Once, Arnu says they dug a sensor up just a few hundred yards from the security guards. "If they've got a problem with this, they'll let us know," thought Arnu. "They'll flash their lights. Get on the horn and yell at us."

When they didn't, the two continued about their work: took the sensor apart, put it back together, buried it where they found it. Like robbers who just want to finger strangers' jewelry and then walk away.

Later, the FBI visited his friend's house, and accused him of stealing a sensor, soon filing a felony charge against him (it was dropped in 2005).

The feds also allegedly called Arnu at work. "My name is so-and-so, FBI Joint Task Force," he recalls the voice saying. "I need to talk to you about your actions in the vicinity of Groom Range."

"Well, I'm at work," Arnu (says he) said.

"Well," (he says) they said, "if you want to have it that way, we can come and have you arrested."

"Why don't we meet after work at my house?" he recalls responding, and then promptly got himself an attorney.

When they met him, and his lawyer, they didn't have a valid search warrant, he claims, "but they made it very clear that we overstepped the line."

"That was the only time when things very quickly got from friendly to 'Whoa, okay, I'm not going to do this anymore,'" says Arnu. By which, of course, he doesn't mean he's not going to do *anything* anymore: He's just going to steer clearer of the line.

The Tahoe moves along the dirt road, and the scanner *beep beeps* a few more times as Arnu describes the shrinkage of public property around Area 51, which continually claims to need more airspace, more land space, and more security. While people could once legally sit on the remote edge of a dry lakebed and whoop while planes took off, that freedom ended in 1984, when 89,000 acres went into the Defense Department's pocket. After that, the edge expanded even farther.

For a long time, there was an on-limits mountain: White Sides, named for the exposed rock cliff wrapping around one face. But in 1995, the military (the base is now an outpost of Edwards Air Force Base, rather than a CIA operation) moved the bar again with the seizure of 3,972 acres and even White Sides became a red zone, along with the ironically named Freedom Ridge. In 2015, the last private land with a view of the area went to the Department of Defense—eminent-domained after the owners refused to turn it over. Now, the view from the outside is much fuzzier, where it exists at all. And Area 51, which began as a 6-by-10-mile patch of land, is almost twice the size of Delaware.

Arnu points out the windshield again, at two objects on a hill: the guards' cars, brand-new Ford Raptors. "At this point, we've hit three sensors," Arnu says. "They know which way we're moving. They know how fast we're going. They're maybe watching us."

When we pull up to the bottom of the hill on which those cars sit, dust rolls up and around our car like a wake. In front of us lies Area 51, although that's hard to believe, and although mountains block us from seeing anything interesting. There's not even a gate to guard the facility, not even a guard shack of the type that protects your local upper-middle-class neighborhood. There are just stop signs, and a big warning that says photography is prohibited, as is trespassing, and also drones. White reflectors run across the road. Behind them, a wire like a heavy-duty Christmas-light cord wraps around a tall tower, at the top of which is a cycloptic camera.

I wave.

The guards turn on their headlights (and perhaps turn down their XM radio) but don't come down to greet us, and no alarms sound when we get out of the car.

It's all very underwhelming.

Perhaps sensing my surprise and disappointment, Arnu cuts in. "As soon as you cross the barrier, things change dramatically," he says. "They get bad very quickly."

Generally, people stay clear and don't get deadly force authorized against themselves. But sometimes people do trespass—either to see what will happen, or by accident. Which is surprisingly easy to do: In many places, the boundary is not even *this* well-marked. It takes the form of spaced-out orange poles, like the ones that edge snowy roads. Unwitting violators of the orange edge can easily end up with their faces in the dirt, eating dust, waiting on the sheriff—just as Wilson described happening to him. Over a loose-rocked hill, which we climb, you can see the more substantial security forces that are just down the road, to deal with people who truly break the rules.

And sometimes people do come here with Intentions. Arnu believes the military monitors the seedier conspiratorial corners of the Internet—websites like *Above Top Secret*—watching for someone who boasts they're about to drive into Area 51 with a bomb. Or, as a meme-creator on Facebook did in 2019, raid the base in a joke that became real.

Despite the bleak picture he's painted so far, Arnu has actually gotten a pretty detailed view of the inside. Anyone can, with a few thousand dollars and the customer-service line of a satellite imaging company. Because, much higher than Tikaboo Peak, orbiting instruments take constant pictures of our planet.

This digital imagery started to become available in 1997, when a company then called EarthWatch launched the satellite EarlyBird-1. An earlier version of the company formed in 1992, when the US government was set to pass the Commercial Remote Sensing Act, which gave private players the right to collect orbital pictures for the first time. Now, that company is called Maxar, and it snaps much of the country's high-resolution satellite imagery (just check the copyright on your most recent Google Maps search). In 2007, Arnu contracted with the company (then called DigitalGlobe) and found that for $20 a square kilometer, and a minimum of a few hundred square kilometers, he could peer inside the base by proxy. He also commissioned images in 2009 and 2011.

When his first data downloaded, he tried to deduce what he could from its pixels, just like satellite intelligence analysts do to figure out if North Korea is moving or testing missiles, or if pirate outposts have popped up on the high seas. He noticed where cars were parked, pegging those as active sites. He saw that new construction had started down south. He found that places he thought were abandoned actually have hangars and hangars of old projects. And, most tellingly, he noticed the way all the hangars hooked up to runways. "You can see the taxi line going in and out of the hangar," he says. "You can kind of guesstimate what kind of projects are in there."

He speculates that the giant B-21, for instance—an early-stage stealth bomber—might live in a giant new hangar with only one center taxi line.

He looks out at the site one more time before we get back in the car. As we drive away, we pass another guard in another Raptor. Arnu puts his hand up and gestures hello to these strangers.

"I always wave at them," he says. "And they don't usually wave back."

In fact, they cover their faces with their forearms so we can't identify them.

We drive back the way we came—down the car-disappearing road, up and over the pass, down into Rachel, and around toward the mountains, headed for the other gate. To our right, farmers have planted fields of alfalfa, shaped like the agricultural circles you see when you fly over the Midwest. But they seem stranger here, alien-green in this arid landscape. And so perhaps it is no surprise that some see these shapely crops and conceive of sinister or savior explanations. To put it more succinctly, as Arnu does, "People think it's a saucer landing spot."

As we approach the gate, a camera watches us from one of those wire-wrapped poles. A small guard shack, not unlike a Swiss gondola, sits near the gate. A metal-sided building, not unlike a warehouse, hangs a little farther back. "If I run up to that fence and jump over the fence," says Arnu, getting out of the car, "all hell is going to break loose."

He points at the road, which winds around the far side of Chalk Mountain. "Area 51 is right there," he says, as if maybe he could reach out and touch it—or better yet, understand it, which we all know is a different thing entirely. He comes out here at night sometimes. He can barely see Rachel. But he can see the radar sites. He can see lights in the sky. It pains his brain.

"I know they're doing something," he says.

And although he'll likely die trying to figure it out, it's the figuring out that's important to him. In this way, he's the same as those who have a more Lazarian conception of Area 51: If they found what they were looking for, that problem would be solved. They would—seeking, always—probably find another mystery. Just as people ride roller coasters and take acid and BASE jump because the world is too tame and sterile, and no tigers are chasing us, people also sometimes feel compelled to chase wonder, because the world doesn't have many surface-level mysteries left. Volcanoes are plate tectonics personified. Illness is microbes messing up the immune system. The sun is just a star. But maybe—outside our overdetermined world—*other* stars have produced beings better than us. And whatever science has done for us here, it doesn't know anything about them there yet.

Science, after all, can't *prove* those beings *aren't* out there. It can't *prove* they're *not* on Earth. It can't even prove they're not on the other side of that gate. That's not how science works. And so you can keep believing, selecting evidence in favor of your own hypothesis.

Arnu is standing next to a package of raw chicken that someone left on the ground so long ago that it's turned a hardened black. He's pontificating on just how *huge* Area 51 is.

I'm only half-listening in real-time, but a couple weeks after I get home, I play the full tape of our interview.

"It's an area bigger than some states are," Arnu's voice says into my headphones, "to fly their planes."

And then a basslike beat begins to pump in the background of the recording. A trebly whine follows, crescendoing till it drowns out his words. The bass and the treble harmonize, then dissonate, then devolve into static, which first drones and then pulsates.

The bass-treble-static pattern happens again a few minutes later, and then again.

It is only when we get in the car, drive away, and head to the crash site of an old F-4D jet that the obstruction stops and our normal human voices return.

Though I know it's just interference between my recorder and whatever radio-emitting systems the site uses, I indulge for a nanosecond the fantasy that there was something They didn't want me to hear.

IV

On the thin road up the mountain, we pass the rancher—Steve, he of the mailbox—driving a giant water tanker, filling up oases for his cattle beyond the Area's boundary. Steve rolls down his window to say hello to his neighbor, and Arnu explains that we're headed up to see the debris from a young pilot in a powerful jet who lost his life against a rock outcropping decades ago.

"I'll let 'em know what you're doing," Steve says, and picks up a radio that connects him directly to those same guards who won't even wave at us. Steve explains to them: Arnu, crash site, some girls.

"Well," says the guard, static crackling behind him, "we still have to follow 'em."

Arnu shrugs—he gets a kick out of this surveillance—and we roll on, up a deeply rutted four-by-four road. This road ends, simply and abruptly, at the top of a hill, surrounded by scrubby bushes. A rock juts like a shark fin into the air.

Arnu turns off the engine and turns to me with a smile.

"Okay," he says. "This is where I charge you how much?"

That night, we sleep at a place called Campfire Hill, just a couple miles from the front gate of Area 51.

A chill comes into the air just as our flames come up, and—sipping wine from our travel coffee cups, eating peanut butter and jelly sandwiches—we feel giddy. We're not afraid of here anymore. Its mystery feels, in some way, like *ours*. Like a light electric touch at the back of the neck. Like an autonomic reaction that twists the mouth's muscles into a smile, tilts the head to the sky, and makes the mouth say, "WHERE ARE THEY?"

What I mean is, I get it. I get why people come here. And I get why people stay.

Just before dark, the guards come screaming up the dirt road. They pull their brakes fast, kicking up dust like a skirt swishing in front of a dancer. Sitting at the end of our road, they shine their lights at us.

We stare at them (or their tinted windshield); they, presumably, stare back. And since I know they cannot bother me unless I'm doing something wrong, which I know I am not, I do not find their gaze intimidating, nor do I find it intimidating to return it. We are simply two different species staring across spacetime, unable to communicate.

The guards rev forward, up to the base of Campfire Hill.

This time, we wave.

They don't respond, and then they leave, and the night lights up as it did before. The giant fleet of tiny white blinkers appears, as if popping out of a wormhole, and floats across the sky in unison. The flares fade in and out of existence.

In having seen them before, in knowing they're not anomalous but, in some way, scheduled, we feel wise. We watch and point, gathering evidence—of a mystery, yes. But one courtesy of our own government, whose power and technology are sometimes so unimaginable, so invasive, and so *per*vasive that humans attribute that might to other beings. In some ways, that attribution feels less scary and terrain-shaking than the truth.

V

In the morning, we stop by the A'Le'Inn for breakfast. We belly up to the bar, where the tender, Dalton, greets us with uncommon exuberance, and immediately sets out to sell us alcohol. Dollar bills hang above the bottles, the top shelf of which isn't quite so.

The two guys next to us smile at their phones while Dalton pitches us the very potent "Cherrya 51" (*So good you won't know you were abducted*). It is 10 A.M. Dalton points at the guys: It's what *they* are sucking down with biscuits and gravy.

Midpitch, Lane pivots: "You wanna get probed?" he asks.

Like a kid who stops conversation to show you his new toy, he pulls down a branch with a huge burl on its end and makes like he's going to come out from behind the bar and stick it in one of us.

"Haha, haha," he says, then proclaims that he's not afraid of stuff like this, like sexually violent alien contact: He's been kicked in the dick by a horse, so after that, you know, how bad could anything be?

He returns the probe to its place and begins to make Cherrya 51s for Bekah and Carolyn while I try to talk to the two men—farmhands whose truck broke down here last night. I explain this book. Hearing me, Dalton, who's been running back and forth between here and the kitchen, slides his hand under a three-ring binder and plops it down in front of me.

"You might want this," he says with a knowing glance.

"Sightings and Reports Book," reads the title. I open the cover, sipping my coffee.

Then, in come two other men, their eyes like saucers, their smiles like cigars. One has a Steven-Tyler mouth and long, shaggy hair. The other looks more classically put together, like a guy who might play golf once a quarter.

"Beer and a shot," says the scraggly guy, before specifying which kind of each he means. They're on their way to a UFO conference in Laughlin, Nevada. Just passing through here, they say, although we all know it's not on the way anywhere, including but not limited to Laughlin.

Guessing he wants it, I slide the sightings book over to him, tell him about my reporting. He flips through the binder: saucers, faster-than-light travel, mutilated cows. News clippings amassed like in a conspiracy shack. He has several shots and a blue drink in the course of an hour.

Dalton encourages this. Dalton is a drinker, drank here at the Little A'Le'Inn for five months, actually, before he became the bartender. He once worked at a nearby farm. "They didn't like my $500 bar tab," he says, laughing. But he liked it just fine, and the boss liked that he already knew many of the customers, so she hired him.

The conversation ranges for a while, then—from "females who can unhinge their jaws," to how Dalton took some UFO tourists/city folk out shooting last night, to how he once made fake wounds on himself to teach his son to deal with outdoor emergencies.

It's only when we get up to leave that the two UFO guys actually begin to talk about UFOs. The topic comes hurtling out of them like water from a burst pipe, like this is their last chance, and they know it.

"Have you ever *seen* something?" they ask.

It is strange, this way of phrasing things, as if only the alien or unexplained counts as *anything* at all.

I mention the flares of last night, the blinkers whose origin we don't know but that were surely just a test.

"So no," I say, "not really. I haven't really seen anything."

Doesn't matter: *They* have seen things—not here, but in life. A cigar-shaped craft, ushered in by jets. A series of lights that appeared and disappeared. Freaky, you know? But the golf guy also once went to an air show at Beale, a "total CIA base." There, he saw jets hover and seem to hop, making boxy patterns in the sky.

Imagine seeing a jet do that, he says, *not* at a total-CIA air show, and in the dark. "It really put UFO sightings in context," he said.

I nod, wondering at the same time if they'll be back here. If they're going to get hooked after this pass-through. And walking back to the car, I recall what Arnu told me, when I asked why he decided to stick around.

"You're going to think it's strange," he said.

"I won't," I said. "I promise."

And so he continued: The first day he spent in Rachel, he burst into the Little A'Le'Inn and shouted a version of the question my new friends saved for the end: "Has anyone seen a UFO?"

He laughs, now, at his greenness, at how he followed such an expected trajectory. Because his decision to hole up here had nothing to do with that—with UFOs, with the inn. It had to do with what happened when he left.

When he drove away from the inn that first night, he stopped the car and just looked at the deep black sky, so bright with stars.

It was that, he says: the desert. The big, unconquered isolation of it.

Out here, you really are alone.

EIGHT

THE MOGULS OF
ROSWELL, NEW MEXICO

I

Not long before Area 51 grew into legend, Nevada's neighboring state—New Mexico—started cultivating its UFO infamy. Today, it's on full blast. As you drive into Roswell a few edge-of-city businesses provide a preview of what's to come when you arrive in downtown proper. A gas station offers an alien discount, whatever that means. Little green 2-D men stand in front of two-story motels. And in July 2018, orange LED construction signs warn drivers that the city is in the midst of its annual UFO festival. Main Street is blocked ahead, in commemoration of that time when a flying object crashed on the nether reaches of a New Mexico ranch, kicking off perhaps the most enduring of all ufological conspiracy theories.

The seminal event—a craft's crash outside of Roswell—took place in the summer of 1947. But it didn't take on its modern alien gloss till decades later. At this point in its transmogrification, whatever actually happened in that postwar heat almost doesn't matter anymore.

Because as the actual crash receded into the rearview, the legend it spawned took on a life of its own.

So, too, did the location. Today, Roswell, population 47,775, fully embraces its extraterrestrial reputation. And why not? Tourists spend more than $90 million here every year, according to a 2015 study, and there's not much to warrant a visit besides the little green men. The UFO festival is the town's premier event, drawing around 15,000 people, who range in enthusiasm from hardcore know-it-alls to casual drop-ins with the barest sketch of the town's history in their heads.

At the 2018 conference, the day is so hot that official fliers warn that dogs should wear shoes or risk blistering their paws. A big stage hosts singer-songwriter-coverer country artists, who serenade spread-out crowds chomping on corn dogs and spiral fries from the food trucks lining the road. A petting zoo sits on the corner that leads to blocks of exhibitor tents. Those tents don't host, for the most part, "the truth is out there; take my pamphlet" vendors but kitschy souvenirs. You can buy a "life-sized" carved wooden alien, like those bear statues people put in front of rural gas stations. Or perhaps you prefer a geode branded as an "alien egg." But one permanent brick-and-mortar store, whose sign says simply "UFO AND ALIEN STUFF," has swag that's more slam dunk on the actual story: a camo-patterned shirt screen printed with the words "Weather Balloon Tracker" and a hat that says, "Rowell Cover-Up Cafe."

Those sayings are on point. Because here in Roswell, ufologists have an actual cover-up. There are decades of governmental lies and obfuscation to point to. And with it, researchers can claim, fairly, that there's no reason to believe that history won't repeat itself.

Everything started on June 14, 1947, at a ranch that's miles from Roswell itself. There, on the flat, prickly land, young Vernon Brazel and his father, William, came across some strange material. They later described it to the *Roswell Daily Record* as "a large area of bright wreckage made up of rubber strips, tinfoil, a rather tough paper, and sticks," not unlike a giant downed kite.

Not thinking too much of it, the family left the debris to bake in the desert sun, so searing it sometimes feels as if it's tattooing itself on your skin. It wasn't until more than two weeks later, on July 4, that William, his wife, Margaret, and his fourteen-year-old daughter, Bessie, went back to that large area and picked up some scattered bits to take back to the house. It was fortuitous timing: Fifteen-hundred miles away, in Washington state, Kenneth Arnold had *also* seen something he didn't understand—those nine lights traveling above the Pacific Northwest's snowy mountainscape, the first modern UFO sighting. Arnold's seemingly impossible saucer story reached William Brazel on July 5, just one day after he brought the desert debris home. It made him think perhaps his debris was not unrelated to Arnold's saucers. "He wondered," the *Roswell Daily Record* reported, "if what he had found might be the remnants of one of these."

Wondering thusly, Brazel traveled into town and "whispered kind of confidential like," that he might have found a busted-up flying disk on his property. The sheriff, suspecting the bundle belonged to the military, took the Brazels' tip to officials at the nearby Roswell Army Airfield. And soon, a major named Jesse Marcel and "a man in plain clothes" went to Brazel's ranch to collect the rest of the bits. The rubber parts were smoky gray, spread over about 200 yards. If you stacked the sticks, foil, paper, tape—some plain Scotch and some with flowers printed on it—into a bundle, it would have been three feet long and seven or eight inches thick. All of it together weighed only about five pounds. No engine, no metal. No words printed anywhere, although there were a few English-alphabet letters, nonsensical on their own.

That disturbing debris in hand, Brazel, Marcel, and normal-clothes guy attempted to Humpty-Dumpty the pieces back together again. "They tried to make a kite out of it," continues the article, "but could not do that."

If it was not a kite, it was a mystery, and if it was a mystery, it must be a flying saucer. Or at least, that seems like how the logic went. Based on this scant information, the Roswell Army Airfield issued a press release that reads like a tabloid:

> The many, many rumors regarding the flying disc became a reality yesterday when the intelligence office of the 509th Bomb Group of the Eighth Air Force, Roswell Army Air Field, was fortunate enough to gain possession of a disc through the cooperation of one of the local ranchers and the sheriff's office of Chaves County. The flying object landed on a ranch near Roswell sometime last week.

It hit the news wire and the local paper, with the heady headline "RAAF Captures Flying Disc."

But that identification wouldn't last long. The debris made its way to the Eighth Army Air Force headquarters in Texas. Staring it down, brigadier general Roger Ramey and his staff saw not something like a kite or a saucer but like a weather balloon.

For sure, the base's weather officer confirmed.

Wanting to set the record straight on this disc debacle—an embarrassment for the military—Ramey invited reporters to come view the debris and see for themselves this was all a misunderstanding. The debris was simply a meteorological instrument—nothing to see here. Mystery solved. Or so it seemed. Or so it was meant to seem.

This, it turns out, wasn't the whole story, and neither was it a true story. But it was, at least, the end of the story for the time being.

It's a wonder that out of that initial incident emerged today's version of the Roswell story—a version in which the name of a town is synonymous with mythological events, and the town itself a place with a UFO festival, on the second day of which there is a pet costume contest.

I see the person in a full spandex green alien suit before my sister, who came with me, sees the Man in Black on stilts. The diversity of life, from the chihuahua wearing a saddle with an alien on top to a lizard with a third eye to a huskie dressed as an astronaut, is astounding. A dalmatian has googly eyes glued to its spots, a UFO hovering over its torso via a stick. An otherworldly goat just keeps eating grass.

"We have all kinds of beings," says the emcee when she steps to the mic.

The shadow from an artillery statue on the edge of the park leans away from her. Participation, she says, is up 40 percent from last year.

II

Toward the middle of the day, when it seems hot enough to melt the pavement, I head to a program whose name sounds like it cannot possibly disappoint: an alien autopsy, a feat performed six times that Saturday at an office supply store called Tascosa. Inside, the audience sits in for-sale desk chairs, printers and clipboards piled up behind them. Strange creatures that do not exist on Earth sit on a table at the center of their circle. These creatures are made not of DNA or RNA or XNA. Instead, they're constructed from food.

The program lead smiles from behind the table. He's an educator from the New Mexico Museum of Space History, located in nearby Alamogordo, the closest town to where scientists tested the first-ever atomic bomb. He looks around at the audience and smiles, about to ask a trick question and a little smug about it.

"How many people think we've found life beyond Earth?" he says.

The crowd—which includes a group of college-aged kids wearing matching homemade T-shirts, old solo men, and families with bouncing children—spins a little in their chairs, looking to see what the others say. As those others begin to raise their hands, so do the other others.

"No," says the educator. "Seriously, we haven't."

People are *looking* for life, he continues—but not in the plains around Roswell. The lookers are astrobiologists, not ufologists, planning searches on the moons of the solar system's giant planets, and in the atmospheres of exoworlds light-years distant. Gesturing downward, he asks the audience to look at the fictional animals laid out before him. They are, he says, optimized for the planets they came from.

"This is a spiny pickleworm," he says, picking up one of the beasts. "These creatures are from Europa."

He holds it aloft, and a couple of little kids stare with wide eyes and open smiles at said pickleworm, a fermented cucumber with a toothpick and feathers stuck in its middle. When the presenter beckons three kids to the operating table, they twist and turn their bodies in the way of shy kids seeking attention.

"Why are its eyes on the sides?" he asks the kids.

At first, they shrug and scuff their feet. But gradually, they wend their way to the idea that this poor pickle is prey, not predator, and so needs to see what's coming up laterally and from behind. The feathers, the kids learn as they pick apart the pickle, are gills—organs that help the worm breathe in Europa's ice-covered ocean.

The whole thing, obviously, is an exercise in evolutionary imagination: If spiny pickleworms *did* exist, what would they be like? And why did they morph in those particular ways? In some respects, these are the same questions anthropologists ask of the Roswell myth, which has also morphed over time, mutating to suit itself to shifting political and cultural environments.

For years, Roswell's now-famous "incident" just disappeared. People didn't have ready access to information—FOIA requests, the

Internet—about the government's shadier dealings like we do today, and so weren't primed, in general, to expect the worst. If Ramey said it was a weather balloon, why would they spend time worrying that it might be anything else? It took decades for Roswell rumors to emerge and, from there, to migrate in from intellectual edge to center. That began happening in the eighties. UFO stories, wrote anthropologists Susan Harding and Kathleen Stewart, were always preoccupied with cover-ups but "conspiracy, cover-up and repression" became central then, perhaps reflecting culture more broadly. Perhaps a similar set of circumstances explains why the AATIP story has taken hold in the past couple of years. The airwaves and alt-headlines are full of conspiratorial accusations, from both the left and the right, in which every group somehow feels robbed of rights and in possession of the true truths behind the surface-level lies.

In this milieu, the Roswell myth started, in many ways, with author and ufologist Stanton Friedman, who passed away in 2019. In 1978, Friedman heard a secondhand story that piqued his interest—one about a scientist who'd seen bodies and a saucer in New Mexico. The pique was still there when Friedman later traveled to Louisiana. There, he spoke to a local air force vet who'd also *seen something* out in the desert, back in the day. That guy turned out to be Major Jesse Marcel, the one who'd gone to collect debris with the rancher. Marcel had never believed the wreckage was a balloon of any sort, no matter what his superiors said.

Friedman, who went on to write his own books, contributed his research to *The Roswell Incident*, by Charles Berlitz and William Moore. It was the first big work since that initial tabloidy press release to suggest the debris had an unknown and sensational origin. The story they told latched on to nuclear fears at the time. Extraterrestrials wanted to keep tabs on humans' scientific progress, gravitating toward atomic and rocket research. Dipping low over Roswell to check in, one saucer was struck by lightning, and pieces of it fell on Brazel's ranch. Though maimed, it continued flying, later crashing on the Plains of Saint Agustin, around 100 miles away (as saucers fly). The military mobilized, thinking it was a crashed plane, but before they could reach the site, archaeologists found it—and its alien passengers. Back at the Brazels', the family found the debris, and the army came up with the weather balloon cover story. End of tale.

More interesting than this book, though, is the similarly titled but extremely different *UFO Crash at Roswell*. In it, anthropologists Benson Saler and Charles Ziegler and physicist Charles Moore analyze the "mythogenesis" of Roswell: how Roswell became what Roswell is, evolving from a small set of details into a fully fleshed creature, in "a process of transfiguration that involved successive retellings in which some of the historically recorded events were retained, some were distorted or repressed, and entirely new elements were inserted."

Back when Brazel first reported this saucer, many people didn't associate saucers with alien spaceships. By the time the first book came out, though, that had changed. The desert was fertile ground for a more celestial interpretation. "The gradual change in the meaning of these terms was paralleled by the creation and growth of a UFO community," write the authors, "[made up of] individuals linked by a common espousal of the extraterrestrial hypothesis as an explanation for some of the thousands of UFO sightings reported over the years."

None of those thousands of sightings, some of them crashes, ever reached the cultural climax that Roswell did. And Roswell didn't reach its own cultural climax till the zeitgeist was ready. The public's affinity for *crashed* saucer stories had waned in the decades between Roswell's incident and the Roswell myth's rise. During those years, ufologists both perpetrated and disproved crash hoaxes, lowering the signal-to-noise ratio. People waved away new reports. Until, that is, ufologist Leonard H. Stringfield suggested that the military semiroutinely hid demolished alien saucers. In this, Stringfield gave the community a precious gift: unfalsifiability. If saucer stories are "not discredited, they are true," the anthropologists write, "and if discredited they are also true, because discrediting is part of a government cover-up of the truth."

From this soil, the Roswell myth rose. The next installment came later in the eighties. In this one, a saucer exploded, sending parts and bodies around Brazel's ranch. The military's cover story was supposed to bolster national security and placate people. By this time, political deception like the Iran-Contra Affair had happened, and the populace knew—because of the Robertson Panel Report—that the government feared UFO sightings could cause chaos. People doubted the benevolence and truthfulness of those in charge. A government cover-up hit home.

Successive versions, of which the anthropologists analyze four, added and subtracted details. But a few ideas remained constant through the years. "Folk narratives stem from, and are supported by, the ideas and beliefs of the subculture," writes Ziegler. "It would be expected, therefore, that subcultural ideas that remain unchanged will appear and reappear in successive versions of the myth."

Roswell's many iterations, and so our subcultural ideas, point to a few stalwarts: The government doesn't want you to know about aliens, saucers are spaceships, saucers crash, aliens look like little humans. Another central motif, according to this analysis, easily emerges from the Roswell story and reappears in most modern UFO stories: a monster (the government) is keeping secret something that could help humans, and the hero (the ufologist) must set the something free. In its elemental form, Roswell is a tale as old as time. But why would people resurrect and perpetuate the Roswell myth? The same reasons, say the anthropologists, that people have always passed sensational stories along: performance, entertainment, attention, escape, money. Books, though not the most lucrative pursuit, lead to paid speaking gigs, television appearances, and thought-leader status. Just ask Tom DeLonge.

This myth also brought UFO believers together as an "us" facing a "them." Whenever the "us" shrinks, and belief dips low, group morale sags. And magically, a mutated myth may pop into existence, perfectly suited to the environmental conditions in which it is born. In the early 2000s, UFO interest dropped from its 1990s, *X-Files* high. Sightings went down; belief in visitation dipped; attention drifted elsewhere. But in the past few years, UFOs have been on the upswing, at the same time as doubt about scientific authority and distrust (and disdain) of political authority. Midupswing comes, now, a new story, about a new Pentagon program, new attempts to keep it covered up, and new declarations that the people and their truth will not be silenced.

III

All of the conspiracy theories around Roswell eventually led the government to respond. New Mexico congressman Steven Schiff asked the Department of Defense to prepare a report on Roswell and make relevant records public. After much parsing, a team released a nearly 1,000-page report in 1995. "Despite its impressive length, the report can be summarized briefly," write the *UFO Crash* authors. "There was no crashed saucer, no alien bodies, no cover-up."

But of course, isn't that what a government that covered things up would say?

The government shakes its head: It was honest, open, this time. "If any of the information discovered was under security classification, it was to be declassified," says the report, "and if active or former Air Force officials had been sworn to a secrecy oath, they were to be freed from to it."

This is the real story, according to the air force: After World War II, the military wanted to detect Soviet missile launches and nuclear tests, the first of which wouldn't take place till 1949. They came up with what they called Project Mogul, on which Charles Moore had worked. Long, tethered herds of thirty or so high-altitude balloons stretched more than 600 feet across the sky, sensors swaying with them. There's no cloaking device for something that conspicuous. People saw them, and teams actually used the inevitable UFO reports to help them track where their balloons were—just as U-2 engineers had perhaps used UFO sightings to see how stealthy their jets were.

The Project Mogul balloons, moreover, had to land somewhere—sometimes where people lived. "These occurrences were typical, leading the recovery crews to describe themselves as Balloonatics," says the report, "due to the predicaments in which the wandering balloons sometimes placed them."

When a Mogul balloon came down outside Roswell, the military sent members of the 509th Composite Group, based at the air field, after it. While those guys had high clearances (they dealt with the nukes), they didn't know about Mogul, which officials kept siloed like so many defense projects, perhaps like some of the UFOs today's navy pilots have seen. The officers in Texas also didn't know about Mogul. According to the report, when the weather officer there saw the debris, he actually laughed. "I walked into the General's office where this supposed flying saucer was lying all over the floor," he recalls, in a sworn statement to the Roswell report committee. "As soon as I saw it, I giggled and asked if that was the flying saucer."

It's a balloon, he told them, and radar targets. He'd seen plenty.

It wasn't, of course. And so when Marcel doubted this interpretation, he wasn't wrong in substance—just in detail. According to interviews in the report, Marcel believed the saucer was a saucer right away. He picked up the sticks, trying to convince his companion that the notations thereon were alien hieroglyphs. But the markings' origins actually came from bootstrapping. When the Mogul team began work, they didn't have adequate supplies. So instead of making the radar targets out of serious material, they ordered some material from a toy company: foil-backed paper like you might use in a craft project, balsa wood slathered with Elmer's-esque glue, nylon twine, brass hardware, "purplish-pink tape with symbols on it," says the report.

A secondary government investigation, released a couple years later (subtitle: "Case Closed"), addressed the discrepancy about the number and location of UFO crash sites—just one near Roswell, versus a second one on the Plains of Saint Agustin—and the pesky problem of the alien bodies. Mogul had involved no bodies, alien or human, but the research team didn't think witnesses were making everything up. They'd seen *something*.

The government's conclusions about that *something*, though, make our brains seem feeble: Witnesses, they determined, had conflated events that had taken place years apart, combining them into one all-encompassing crash. When people spoke of alien bodies, they were actually describing multiple events, all separate from the Mogul incident: the battered remains of crash dummies that the Air Force strapped into balloons in the years after the Roswell Incident, *and* the transportation of actual injured human pilots to the Roswell Army Air Field Hospital. "Persons who have come forward and provided their names and made claims may have, in good faith but in the 'fog of time,' misinterpreted past events," said the first report, which—implies the second—was correct.

Ask any Roswell adherent what they think of that explanation, and you'll hear some version of "hogwash." Even the grounded Greenewald thinks this is bullshit. But if you've ever hashed out childhood memories at family gatherings, you'll see that your sister thinks you saw that porcupine on the 1992 Grand Canyon trip, while you think it was the '94 trip to the Rocky Mountains to cut down a real Christmas tree.

The preparers recognize that people are going to say "hogwash," that their thousand-plus pages of work likely won't convince anyone who isn't convinced already. "It is assumed that pro-UFO groups will strongly object to the attached report and denounce it as either shortsighted or a continuation of the 'cover-up' conspiracy," reads a cover letter from Richard Weaver, director of security and special program oversight at the air force. "Nevertheless, the attached report is a good faith effort and the first time any agency of the government has positively responded officially to the ever-escalating claims surrounding the Roswell matter."

That's nice. But it can be difficult to trust a government's "good faith effort" when that very effort demonstrated that the government *did* lie, about these very incidents, for decades. Carrion, for instance, spent more than five hundred highly footnoted pages in his book *The Roswell Deception* laying out evidence that the events around Roswell were part of a "strategic deception": The United States wanted to convince the Soviets that it had technology it didn't have, and to draw Soviet spies from their hiding spots. It's a compelling idea, one that helps explain why the press officer would have cried "flying saucer" in the first place.

Whatever you make of the debacle, the secrets the government has kept in the past make it hard to believe what officials say now. It is like the Roswell anthropologists said: "[If] not discredited, the [stories] are true, and if discredited they are also true, because discrediting is part of a government cover-up of the truth." The conspiracy about that cover-up is, in many ways, due to the government's own faulty response, and it has strengthened some theorists' grips on alternative, alien interpretations. If the government had told the truth initially, maybe no one would have grabbed on to these particular events in the first place.

The tide, today, is turning in a different direction: Those promoting AATIP and the UFO videos gain authority by claiming the government is endorsing them. The government isn't exactly doing that, but the promoters' angle is clear: The overlords don't want to hide the truth from you anymore, like they used to in the bad old days. And we're helping them get that truth out there.

On Saturday afternoon, after the "alien" autopsy, I push past a line of people waiting to enter the Roswell UFO Museum. I slide through the gift shop, up to the entryway of an elevator whose appearance screams "slow." The paid-attendee-only events take place through this portal. Only the serious UFO Festival attendees are allowed up here—no one who's just in town for the T-shirts and corn dogs.

Frank Kimbler's talk, "Roswell UFO Crash, Best Physical Evidence," has a larger audience than most—perhaps because it purports to deal in the concrete, with science and with more than mere memory. Outside the room, the overflow crowd opens the shades on the glass wall and peers inside. Kimber, a small man who looks like he might set out on a birding expedition at any minute, steps behind a podium and pulls up his slides.

"We have to take the fantasy and interweave it with the facts," Kimbler says, "to kind of figure out what the truth is."

Kimbler goes on to describe his investigations of the crash site—from analyzing satellite images to searching for UFO metal fragments.

"Did the military get it all?" he asks the audience early in the talk. "And did they cover up all the proof?"

On the screen, he pulls up images of little bits of metal he found, some buried under dirt. They resemble melted beer can remnants, resolidified. If you've ever looked at the floor of a desert, you know that all manner of human objects ends up there: covered by lonely wind, preserved by dry conditions, left alone. It's not unusual to find debris even in remote stretches. Kimbler, though, believed the metal he found could have been left over from the saucer mishap. And so he picked up some of the pieces, took them back to civilization, and analyzed their composition—which he thought was, perhaps, otherworldly in some way.

"I hope I'm not boring you with science," he says, smiling. "We've got lots of science."

One piece of science: one of Kimbler's samples of aluminum metal contains 3 percent molybdenum.

"It is a no can do on this planet," Kimbler proclaims. He says that if you mix molybdenum and aluminum, they will explode. This is all an appeal to scientific authority (when science supports ufology, it's suddenly OK, just as is true in climate denier and antivaccine communities), a request not to question. Nevertheless, I pull my phone out of my backpack and open a browser—because, by definition, nothing that exists can be impossible. After the page loads, I see that—lo, behold—the first Google result will sell you metal that combines those two elements.

"If you do good science," Kimbler says, as I put my phone away, "the debunkers can go jump in a lake."

At the end, a wide-eyed spectator raises a hand. "Do you think the pieces you found are from the interior or the exterior of a ship?" they ask.

"Yes," says Kimbler.

IV

That night, in the dingy Motel 6 a few miles from the festival, my sister and I turn on the TV, where the movie *E.T.* is coincidentally about to start.

"Let's watch it," I say, thinking that watching this movie while at the Roswell UFO Festival will make a great scene in this book. Whenever my sister tries to make jokes or talk about present-day Drew Barrymore, I inform her that I'm "trying to have an *experience*."

In case you've forgotten the plot of *E.T.*, it goes like this: A spaceship crashes to Earth, leaving a baby hippoesque occupant to roam our strange planet, which is full of taller, more graceful, and decidedly less empathic beings. A young boy, Elliott, discovers E.T. and hides him from all the adults in his life.

Elliott and the alien become physically and psychically connected. The thoughts and health of one affect those of the other. If you're scouring a film for meaning, this could reveal our desire to relate and be relatable to the alien. If you're a kid, you mostly consider the scene where E.T. drinks a bunch of beer, and, at school, Elliott becomes drunk and lets a bunch of frogs free from the dissection table.

At the climax of the movie, various adults—from The Government—attempt to take E.T. away from Elliot. The Adults wear sterile suits, faces covered. They barely look human and act like an unspeaking horde, not differentiable individuals. This is The Government at its most impersonal, its most overreaching, and, really, its most alien. As this intrusion is in progress, E.T.'s health (and so, too, Elliott's) declines.

When the agents snatch E.T. away, Elliott protests. "He came to me," he pleads with them, desperate. "He came to me!"

He was chosen, special. And he alone could save this alien who was also saving him. And, indeed, he does: Soon, Elliott and his band of mischievous friends escape the human horde, stealing E.T. away to his waiting spaceship. Their bikes fly over their town, over the authorities, over the whole Earth, powered by E.T. himself, who has made them part of the sky. Finally, they land hard in the forest. Before them sits a real spaceship, lights on its equator looking like an arcade game.

"You could be happy here," Elliott had previously told E.T. ". . . We could grow up together."

"Come," E.T. replies.

"Stay," responds Elliott.

E.T. rests his glowing finger against Elliott's little-boy chest, where his fragile, human heart beats.

"Ouch," says E.T.

"Ouch," says Elliott, tears sliding down his cheek.

E.T. stretches his finger toward Elliott's head, touching his face but meaning his mind.

"I'll be," says E.T., "right here."

The lesson of the movie is clear: People may not believe you when you say what you've seen. Powerful entities may try to take your experience away. But it's real, it's yours, you're right, you're special. It will be right here, no matter what.

"Did you know Drew Barrymore once spray-painted her ex-boyfriend's car?" my sister asks.

V

Media-based fictions exist about Roswell, too, invented whole cloth just like E.T. Curt Collins, who runs the website Blue Blurry Lines, knows this personally.

"I'm like a UFO Rip Van Winkle," Collins says when I call him after the festival to talk about one in particular. He was interested in UFOs as a kid, but like so many typical childhood obsessions—dinosaurs, Legos, becoming an astronaut—UFOs crashed by the wayside when he aged. The interest reengineered itself, though, when Collins watched a documentary called *Shades of Gray*, about a UFO icon named Gray Barker. Barker wrote the book that first introduced the world to the Men in Black, those guys who arrive in suits to erase your memory if you get too close to the truth.

On the day-to-day level, Barker seemed to take UFO claims at face value. "If there was someone who claimed to have ridden on a flying saucer, he'd write articles about it," says Collins. But he also perpetrated hoaxes and took considerable literary license, and that dissonance attracted Collins: Sure, the guy was peddling cosmic snake oil and spooky bedtime stories, appealing to audiences as much as possible. He was also skeptical of the field, but at the same time he appeared to be generally interested in it. Or was it just about the money? "How could someone have these mutually conflicting ideas about a topic?" says Collins.

Soon, delving into research, Collins learned that many hoaxers are also believers: They want evidence so bad that they're willing to fabricate it.

Amidst those fakes, though, Collins sees what he calls "genuine scientific mysteries." "It's those that inspire me to keep looking at the subject," he says.

In 2013, Collins was still a relative newcomer, but he started his own discussion group—private, full of people who were not afraid to call bullshit on ufology's boundless excitement. That structure would soon allow the members to help take down a flashy hoax—one that reflected the ones that came before it and flashed forward to those that would come after.

It all began, allegedly, in 2008, when a woman named Catherine Beason was cleaning a to-be-demolished house. She grabbed a box of photo slides, and, years later, noticed two strange snapshots, both of some kind of tiny corpse in a glass case. She passed them along to her brother. They reminded him of the stories about Roswell, of the short, big-headed beings that allegedly died there. But he didn't do anything about that reminder until 2012, when he passed them along to ufologists Don Schmitt and Tom Carey, who wrote some of the Roswell books.

When Carey saw the pictures, he said he "knew" they were genuine. Later that year, Beason recruited a filmmaker to chronicle the investigation of the images. And in 2013, the group formed the idea for "BeWitness," a three-ring show at which they would reveal the pictures to the public. By then, despite nondisclosure agreements, rumors had begun to circulate among ufologists about the pictures. But no one knew anything for certain.

"We were really starved for details," says Etienne Hamelin, a French UFO researcher who belonged to Collins's group. "Having no access to the slides, we had to wait for information to be released, indirectly, on blogs, as everyone else."

The break came in November 2014, when Carey gave a lecture at American University and tantalized the audience with a teaser of BeWitness.

"We have a lot more," Carey told the audience. "We're still working on the case, but this is big stuff."

At the official announcement a few months later, Schmitt spoke with grandiosity. "It will certainly be the most important event in our lifetimes," he said, as the team played a trailer for the investigative documentary, *Kodachrome*. In the footage, most shots of the body—presumed dead, presumed alien—were blurred to delay the big reveal. But one short segment had a clear picture of the picture. Collins and his colleagues pounced.

The photo was still blurry and stretched-out, bereft of the detail beyond the obvious facts: It showed a small, strange body, mouth gaped open like a ghost, chest caved, limbs too small. The disproportion was unnerving, says Collins. Unearthly.

This photo, secondhand from a blue-tinted screen, *did* look kind of alien. But Collins thought the corpse's resting place was all wrong. Curators of a dead extraterrestrial would have protected their specimen with steel. "That did not look like a top-flight museum case, and it didn't look like a military lab," he says. "That was one of the things—why would you put a precious interplanetary specimen on a glass shelf for display?"

"Of course," he adds, "this didn't prevent the ufologists from building complex scenarios of a discovery in a secret lab. But it was all make-believe."

Collins, Hamelin, and the others began trying to unravel the mystery and the story. They searched for visual matches among "human oddities"—the kinds of things you'd find at old-school circuses and freak shows, like mummies, "mermaids," and the like. At first, they just discussed it amongst themselves, and sometimes in other UFO Facebook groups, like Collins's. But there, algorithms kept sliding comments about the Roswell slides down, so one member suggested they create a new online corner solely for discussing the slides. Within that oasis, one researcher found a child-mummy—affectionately called 2397 after its Smithsonian label—that resembled the "alien." They unearthed a paper an anthropologist had written about it, and called him up to get his opinion (it looked, he thought, strikingly similar).

"There was sort of a competitive spirit to see who could produce something," says Collins. "To see if we could find something to match. It sort of fed on itself."

One day, a member created a website for their Facebook group, dubbing them the "Roswell Slides Research Group."

"All of a sudden now there's these expectations of us," says Collins, "and we think, 'I guess we better do something.'"

They began to publish papers about their doubts, and researcher Tim Printy put his results in his well-read SUNLite newsletter. But the big break came from the BeWitness event, to which people bought $20–$80 tickets and paid $20 to livestream. Seven thousand people came, and 2,000 people tuned in.

Despite the sticker shock, the show started out anticlimactically. "I remember vividly," says Printy. "Most of us got bored . . . When they finally showed the slide just before intermission—because there was intermission, that was how long it was—I was like, 'Is that it? Is this what we paid for?'"

The slides appeared only briefly, the big focus being on CGI animations of the alien face fleshing itself out, a hologram showing what its body may have looked like. This being was not, the presenters said, even a mammal. Maybe it used electromagnetic radiation to communicate.

OK, thought the Roswell Slides Research Group. But the (five-hour!) presentation did yield a breakthrough: better pictures of the pictures—good enough to tell right away that they did not, at least, show the number 2397. In one, the researchers could see that the glass case had a *placard*. With *words* on it. Kind of.

"It looked like a mess to me," says Collins.

Not long after, a researcher obtained a leaked higher res copy of the slide that showed the placard. He posted it to the Facebook group. Right away, Hamelin started to sharpen the placard with a program called Smart Deblur (not unlike the "Enhance! Enhance!" feature on forensics shows).

The first line popped out easily: "MUMMIFIED BODY OF TWO YEAR OLD BOY."

The second and third took a day or so, but soon, they could read all the text: "At the time of burial the body was clothed in a slip-over cotton shirt. Burial wrappings consisted of three small cotton blankets. Loaned by Mr. S. L. Palmer, San Francisco, California."

"I just can't understand how they were fooled by this," says Collins, sounding like a disappointed parent.

But the revelation didn't dethrone the BeWitness players. In ufology, hoaxers and those fooled by bad data often get to stay in the community, untouchable in their own view and often in that of believers. It's the interpersonal version of unfalsifiability: When the Research Group came out with their assessment, the hoax-peddlers initially said, "You're lying. *You* faked *your* photo."

"'Oh, I shot you,'" imitates Collins. "'No, I used a force field,' 'I used a laser,' 'I have a mirror to reflect it.'"

Eventually, though, after authenticated versions of the photos spat out the same sentences through SmartDeblur, the BeWitnesses had to concede. Printy looks to the movie *Patton* for answers about why hoaxers and embellishers hoax and embellish, why alleged witnesses go beyond the bare facts of an event. At the end of one speech, Patton sets a scene: You're at the fireside, and your grandson asks what you did in World War II.

"You can tell him that you fought in the Third Army, and you won't have to tell him that you shoveled crap in Louisiana," says Printy. That's not true for everyone. "You got these old gentlemen in their 70s, 80s, 90s, and you ask them what they did in Roswell in 1947. What most of them were doing was cooking, working on trucks, flying airplanes, going through the usual exercise routine, and that's all they did. Day in, day out, day in, day out. . . . Some people exaggerate, confabulate, make up stories because, let's face it, I can't tell you about the time I changed the tires on a truck."

The UFO community, to Collins's dismay, keeps learning the same logical lessons over and over. In an essay for *UFOs: Reframing the Debate*, he explains what ufologists could have (should have) taken away from the Roswell Slides. One, demand better evidence. Investigators so rarely get raw data for any UFO case. The people who have that data simply present their interpretation rather than the evidence itself. The same thing, to be fair, also happens in more traditional science: Rarely do other scientists wade through your spreadsheets and refashion your Python scripts to see if their conclusions align with yours. People just accept the inferences of others, without demanding the background. But if another researcher asks, and they withhold, that's fishy.

The same, in the present day, rings true of AATIP. Elizondo and his crew could have given us the radar returns, the whole tape, the original files, proof of employment. They could have given us documentation of the program's initiation, rather than their version of how it came to be, to prove to the public that the Department of Defense intended it as a UFO program. "We just don't know how much of this is what the original program even wanted," says Collins, "Was this tangential stuff they stuck in the file?"

The parallels between AATIP and the Roswell Slides continue. "So what we have here, rather than slides, we have the existence of this program," explains Collins. If we have this slide, UFOs are real and crashed in Roswell. And for AATIP, "the government is studying UFOs, therefore UFOs are real."

But of *course* UFOs are real, and of course the government studies them. They are just, says Collins, anything unidentified in the sky. They are not necessarily alien ships. "The word 'unknown,'" says Collins, "that fires the imagination of people who believe UFOs absolutely represent extraterrestrial spacecraft. Anything unknown to them is an endorsement of alien life."

Culture bakes that connection between UFOs and aliens into us from the beginning. Most people learn about UFOs through entertainment: *E.T.*, *Independence Day*, *War of the Worlds*.

"That becomes their working model," says Collins. "The message we receive is different from the signal."

It's a lesson present-day Roswell enthusiasts have still not assimilated.

At the festival, the last talk I attend is by Stephen Bassett. Bassett often wears a black shirt with a tiny turtleneck as he agitates for official disclosure about alien presence on Earth. He's a better speaker than most—a performer, a preacher, head of the Paradigm Research Group. And perhaps because of this rhetorical skill, people crowd the room.

When he steps up to the mic for his talk—titled "Extraordinary Exopolitical Developments (1993–2017)"—he requests that the lights be turned off.

After darkness descends, Bassett points a flashlight up at himself. A shadow, an ethereal man in black, falls behind his shoulders.

"It's Sunday," he says to the congregated people. "You should be in church."

They laugh.

Depending on how you depart from the Roswell UFO Festival, you might pass by another reason people trust ufologists like Bassett (and Kimbler and Carey and Schmitt) more than they have historically trusted the government. Just 140 miles away, in White Sands Missile Range, engineers performed the Trinity test, the first-ever detonation of an atomic bomb, without much concern for nearby residents.

Twice a year, the army opens the Trinity site—now only radioactive enough that you wouldn't want to stand in its crater all day every day—to tourists. Visitors wait in hours-long lines of cars that put Disney queues to shame. They show their IDs to security guards at the base's gate. They drive past miles and miles of nearly empty desert. Low buildings and high towers and barely marked side roads fly by.

Miles past the gate, the Trinity site has a festival atmosphere during its "open house." There are parking attendants, T-shirts, hot dogs for sale. Viewers get out of their cars and walk the five minutes to the actual spot where, in 1945, engineers detonated a device full of plutonium, bombing us into the nuclear age with the only type of weapon—so far—that could make our whole planet uninhabitable.

There's barely a dent in the ground now to commemorate that terrestrial turning point. And where there was once a bunch of trinitite—the radioactive mineral that fission-melted sand fused into—there is now mostly regular sand. Collectors have bought the trinitite, scientists have borrowed it, and tourists have pocketed it while pretending to tie their shoes.

Along the back fence, the army has placed photographs of the bomb at various stages of explosion. The picture from fifteen seconds after detonation shows a classic mushroom cloud. It is a bomb, definitely.

But if you walk backward along the fence, rewinding the time-line, you'll eventually get to the image taken 0.006 seconds after the explosion.

There, the cloud looks almost like one of those indoor tennis bubbles. Innocuous. Like nothing to be afraid of. Like nothing ever happened here. Like there is nothing, really, to see.

NINE

TO THE STARS ABOVE

TELESCOPE TOWNS

I

I f you wanted to watch the birth of brand-new myth about alien contact, all you needed to do was turn on the September 7, 2018, edition of KVIA nightly news. There, anchors laid out a mystery more confusing than what city councilors in Las Cruces were doing, or what crimes had gone down in El Paso. At the desk in the studio, anchor Estela Casas furrowed her brows and squinted her eyes in suspicion as she described an evacuation in a tiny mountain town—Sunspot, New Mexico, home to a facility that studies the sun.

"The Sunspot Observatory is empty," Casas said. "And it's unclear why."

She cut over to correspondent Mauricio Casillas, who had spent the day at the solar telescope site, about 15 miles from the small town of Cloudcroft, New Mexico—searching for answers.

Unfortunately, Casillas hadn't found many.

"We know that the FBI is the agency that evacuated the facility," Casillas said, "but no one seems to know what exactly is going on."

Behind Casillas, officials had taped off the entrance to the observatory with yellow caution ribbon. No one, including the employees of the post office located inside, could enter. "Even local law enforcement doesn't know what's going on," Casillas continued. "They asked *us* if we knew anything."

It took national news outlets a few days to latch on to this strange story spilling out of rural New Mexico. But others were faster: Almost as soon as the local reporters like KVIA's spoke about the situation, Conspiracy Internet was on it.

And of *course* Conspiracy Internet was on it. The FBI . . . in a telescope town . . . with an evacuation order: It's like an end-of-the-world Clue game.

The day after KVIA's broadcast, a user called SeaWorthy started a Sunspot thread on a conspiracy forum called Above Top Secret. Many parts of this forum are—like most conspiracy websites—full of hate and fear and bad information. But among those baddies and bigots, there are also smart digital denizens. Even if their favored topics are Out There, they keep investigating genuine mysteries long after the general public has let them disappear, like whack-a-moles, from with the news cycle.

"Easy to make myself a disaster movie scenario from the Post office being closed too," SeaWorthy wrote. ". . . If it were a movie the FBI would soon come up with a cover story to put out."

A user called DBCowboy chimed in: "In 'Close Encounters' they staged a mock chemical spill and evacuated a town."

"How far is it from Roswell?" someone joked.

In the r/Conspiracy forum on the discussion site reddit, users also quickly chimed in with cosmic options: Maybe the telescopes atop Sacramento Peak had seen a solar flare that would kill us all. "I hope it's friendly aliens with cool drugs and not a rain of asteroids," said cosmicmailman.

Occasional jests like these suggested aliens or UFOs or alien UFOs. But mostly, conspiratorial types stuck to earthly or cosmically nonbiological suggestions for the FBI presence and population evacuation: a leak from the ten tons of mercury on which one astronomical instrument floats; a spy peering down on nearby White Sands Missile Range, the largest military facility in the US; a threat of arson; installation of secret equipment; biohazards sent via post. Regardless, though, of whether conspiracies are of the celestial or terrestrial type, they sprout from the same root: They represent secrets the powerful know that you don't. In that way, it's always aliens, even when it's not.

Despite the largely just-a-joke nature of the actual conspiracists' alien mentions, when the story got national, mainstream coverage, ET dominated the headlines. "Mysterious Observatory Evacuation Stirs Alien Conspiracy Theories," "Sheriff Describes Mysterious FBI Tests amid Alien Speculation," "Alien Conspiracy Theories Swirl after FBI Evacuates Space Observatory."

Given the clamor of the rumors, or—more accurately, the clamor about the clamor of rumors—an observatory spokesperson even felt compelled to say, "I can tell you it definitely wasn't aliens." She would say nothing else substantial.

The alien theory seemed almost to have come from the headlines themselves, which first presented and then refuted it. Clickbait fodder surely played into this. But on a deeper level, headlines are designed to tap into what human readers most want, fear, and expect. "It's aliens, they're here" represents all of the above. Sunspot's saga is the stuff of feature films. Films in which the world falls apart or we save ourselves or both. Films in which our lives become more exciting or terrible or both.

This is it! we allow ourselves to think, reading about the tight-lipped closure of a telescope town—the town that saw too much. *This is how it begins*. "It" represents the exciting, terrible future, full of salvation and destruction, that we've been told—by religious leaders and movies—is ours to inherit.

The night of September 12, I arrived home in Denver—about nine hours from Sunspot—and siphoned up the stories that I'd missed while away on a trip.

It's so close, I thought. *I could just go see.* Because it was a federal observatory, I knew Sunspot was likely located on federal land—so even if the observatory itself was blocked off, I could probably camp pretty close, and maybe hike closer, without breaking the blockade.

By noon, I had stuffed my backpack and mapped out a base camp for myself, as well as a route to the far side of the observatory. I would try to talk to some townsfolk. I would have polite conversations with FBI agents. I would see whether alien seekers had descended on this small, solar-centric place. I would find out the answers, which I wanted as much as any Above Top Secret detective (by this point, the pages of discussion ran into the dozens).

Driving deep down Colorado and New Mexico's empty highways, I finally arrived in the outdoor-retail-centric town of Cloudcroft, the last outpost before Sunspot. I stopped at the local gas station for firewood. And after I paid, as I was walking out the door, the clerk called to me.

"Watch out for deer and elk," he said. "If it's dark out there, it's dangerous."

The words felt ominous. And it *was* dark. By the time I got to a trailhead called Cathey Vista, the air was black in the way it can only be in rural places: like light has never shone here and may never again. Like just beyond the beam of your head lamp is a monster. An alien, a mothership, Bigfoot, a serial killer. When knowledge hides beyond our senses, we tend to imagine the worst or most powerful or both. Especially, perhaps, when we are camping near a place that has been evacuated for unknown reasons.

I set my spot up a few yards down the trail, squinting to see behind every tree. Hearing aircraft and wondering if they could see me. Listening to the bugle calls of mating elk—which sound like autotuned cries from cryptozoological creatures—and feeling like they would soon be, if they were not already, upon me.

II

O bservatories are a natural place for conspiracies and cover-ups to roost, whether they be of the alien sort or not. These ideas, after all, spring from a general distrust of authority, from a sense that officials are hiding information, and from the historical truth that, yeah, the people with power wheel secret dealings and bury information that should be public. Sometimes those powerful people are FBI agents. Sometimes they are scientists. Sunspot provided the perfect overlap. Here, a special group collects data on the cosmos, in a way inaccessible to the average person, and now the feds had descended and weren't talking. Were they hiding evidence of aliens or their craft? That's a slightly ridiculous notion, but the way astronomers tend to treat ufological claims doesn't help dispel them—here or in general.

Scientists often dismiss UFO sightings and the idea that aliens are already here. Dealing with these claims is not worth their time—because the evidence isn't solid enough. If they don't dismiss, they frequently debunk—because the evidence is only worth knocking down, without any validation or acknowledgement of a person's feelings and experiences. Studies of how scientists and communicators deal with climate change deniers and antivaccination activists show that such debunking isn't very effective, that implying believers are dumb and providing piles of evidence don't usually change minds. The tactics come off as both condescending and as protesting too much—the data-centric version of "nothing to see here."

Communications professor Adam Dodd, from the University of Queensland, has said that astronomers' public declarations about

UFOs are "performative." They treat the topic the way Science expects them to, so that the community continues to accept them. Their words bolster the existing boundaries of science, keeping them inside and UFOs out, banging on the border wall.

In reality, Dodd contended, their statements actually represent "a departure from the principles of sagacity, objectivity, and curiosity that are usually seen as characterizing their profession." For example, he points to a time when famed physicist Stephen Hawking committed a logical fallacy he would have found unpalatable elsewhere: Citing the absence of evidence ("I've never seen a credible UFO or Earth-dwelling alien report") for evidence of absence ("Therefore they cannot be here"). On top of that, Dodd continued, citing psychologist Stuart Appelle, ufology "is not simply rejected as a legitimate discipline, it is categorically dismissed." Those are not the same action: "Rejection suggests a conclusion based on close examination and careful reflection. Dismissal is an *a priori* judgment that close examination is not warranted."

If anyone has ever dismissed you, you know how it makes you feel: First, bad about yourself. Then, you feel that, no, they are the ones who are wrong, and you'd like to prove it. And finally, if you're a bit paranoid, like they *know* they're wrong and have an interest in hiding the truth.

With this mutual antagonism in place, the sites where scientists construct cosmic knowledge—telescopes—provide fertile ground for conspiracies, celestial or otherwise. Add in some FBI agents, an apparent gag order, and a few displaced residents, and you've got yourself the perfect environment for imagination to evolve its own story.

There's another ingredient in the Sunspot recipe, though: Astronomical facilities, including the one on Sacramento Peak, often grow out of military investment. The military is the ultimate big-G

Government, at least to the conspiratorial mind. Plop the two down on the same peak and point them at the sky, and conspiracy will come calling.

During World War II, the Department of Defense understood the upper atmosphere—subject to the sun's whims—played a critical role in their efforts: It affected and responded to radio communication, flying missiles, and supersonic planes. In the beginning, after this realization, the government gathered solar observations from a place called the High Altitude Observatory. Located at over 11,000 feet in elevation on the Continental Divide in Climax, Colorado, this observatory looked like someone threw a barn on top of an office building. Astronomer Walter Orr Roberts was the lone scientist at the observatory most of the time, living in a small ranch house with his wife, Janet, often surrounded by piles of snow. Roberts, perhaps staring at this white blankness, noted that "there were long periods of cloudiness, especially during the winter, when it was not possible to make observations," according to an account from historian Ruth Liebowitz. Because of that opacity, the air force decided to set up a second observatory.

Scientists began searching for a place that would stay sunny when Climax was dark. You know, a desert. After considering New Mexico's White Sands Proving Ground, where soldiers fired off rockets and also the first atomic bomb; next-door Holloman Air Force Base; and Alamogordo, they finally identified a place above all of those: a mountain called Sacramento Peak, 9,253 feet above sea level.

They tapped a young guy named Rudy Cook to be this potential observatory's prime resident. He lived in what would become Sunspot alone for months, holed up in a railroad box car, with only his dog, Rocky, for company. After he'd stayed long enough to report good, stable conditions, the overlords officially decided this was a prime location. So Cook, along with a newly arrived family, installed a Sears and Roebuck garage, a water tank, a bathroom, and a garbage pit. The air force donated a trailer with a prefab kitchen that included a warm stove. In 1948, the air force brought in a dorm that could fit around thirteen people. Eventually, modest houses, an electrical plant, sewer lines, and radio communications with Holloman came about. By the early to mid-1950s, the observatory had houses and a helicopter pad. 'Twas a town.

Soon, too, came the telescopes for which all this infrastructure existed in the first place—instruments that could look at the prominences arcing from the sun, the sunspots that darkened its face, or the flares it sometimes shot out. Holloman Air Force Base, conveniently, also perched equipment onsite to track its missiles and rockets. According to a history in the National Solar Observatory's archives, Sacramento Peak Observatory soon became "the Air Force's leading center of solar studies."

Sunspot did the military's solar work for decades. Only later, in the 1970s, did it transition into the National Solar Observatory, a civilian facility funded by the National Science Foundation. And even then, it continued to do some military work. This is not an unusual astronomical trajectory: The technology that humans use to understand the universe beyond this planet has always been useful to those who wage war and gather intelligence. You can look at the sun to understand the sun; you can also look at it to understand how your commands to soldiers might get scrambled. You can use astronomical tools to track asteroid activity; you can also use them to trace a missile's path. You can use an orbiting telescope to survey the cosmos's most distant galaxies; you can turn its mirror toward the ground and see the movements of earthlings. You can send radar beams crashing into the surface of Mercury; you can shoot them up against an enemy aircraft. You can track the pings from the Voyager spacecraft as it leaves the solar system; you can track the pings from Russia's spy satellites. It's all essentially the same equipment, just facing a different direction.

Astronomers don't like to talk much about the military applications of their work, often preferring to think of themselves as gifting pure knowledge to the species. But it's undeniable that their technology both comes from and pushes forward governments' very terrestrial interests.

So it's not crazy to come up with conspiracies about the sinister work that might happen at observatories. It's not wildly off base to conjecture that some studies stay in the shadows. It's not illogical to see science-military collusion simmering below the surface. History proves it, in some cases, very logical: One of the first SETI (Search for Extraterrestrial Intelligence) instruments, for instance, was actually used in a secret operation—unknown even to some of the scientists who built it—to spy on Russian spacecraft.

Conspiracies swirling around the September 2018 events at Sunspot Observatory, then, parallel the funniest jokes: They contain an element of truth.

Was the government hiding evidence of aliens?

(Almost) certainly not, no.

Was the government hiding something?

Absolutely, yes.

III

The telescopes you probably first think of when you think of aliens—radio telescopes, which are the telescopes that have historically done SETI work—also came directly out of the military. Radio telescope technology only became widely available after World War II engineers had developed radar. Radar equipment broadcasts radio waves that bounce off whatever they encounter and return changed by their experience. Their shape, strength, and frequency tell engineers about the objects they slammed into at the speed of light. Those altered waves end up at antennas, which detect them and turn them into signals that human beings can sense.

Once the military had matured this technology, astronomers—some of whom had worked on said technology themselves—could better appropriate it for their own purposes, because the antenna part of radar works essentially like radio telescopes—except that telescopes pick up radio waves from deep space. Every radio telescope owes its origins to the military-industrial-scientific complex of the mid-twentieth century.

Sometimes, though, the link is even more direct, and a radio telescope—like the National Solar Observatory—begins life as a military project. Take Arecibo Observatory in Puerto Rico. It's a giant dish, 305 meters across, plunked in a natural depression in one of the island's jungles. You've perhaps seen it in the movie *Contact* (which features aliens, the military, and a conspiracy courtesy of a billionaire), or in the James Bond flick *Golden Eye* (which features spies and Russians). Arecibo was born out of the Pentagon's Advanced Research Project Agency, or ARPA. ARPA wanted to detect and knock out incoming ballistic missiles. Some studies suggested that if a nuclear warhead launched through the atmosphere, did a ballistic turnabout in space, and sped back down, it would create a unique atmospheric signature as it—hot, fast—stripped electrons from molecules way up there. But problematically, ARPA didn't know enough about the atmosphere to know what that signature would look like. As part of Project Defender, it commissioned Arecibo's construction to find out and to do straight science.

Construction started in 1960 and ended with the iconic dish in 1963. Alongside, the telescope had a giant radar that could shoot into the atmosphere. The back-traveling waves would tell the scientists about its contents and physics, partly so they could imagine a missile sliding through. It was called the Department of Defense Ionospheric Research Facility, although it didn't do classified work and its research wasn't directly military.

In the 1970s, the National Science Foundation took the telescope over, NASA shouldered primary radar responsibility, and Arecibo became a civilian-focused observatory. After that, Arecibo got right to its extraterrestrial work. In 1974, astronomers Frank Drake and Carl Sagan, among others, designed a radio broadcast that could, perhaps, be understood by beings that knew neither English nor about Earth. In binary code, the scientists encoded a graphical message meant to be a sort of self portrait of human life. It's made of ones and zeros, forming pixels like the bulbs in a Lite-Brite.

The message begins, at the top, by teaching the aliens to count, showing them representations of the numbers one through nine. Below that, it provides information about our personal chemistry. Then comes a sketch of a human and a tally of the number of that species on Earth. This is followed by a line map of the planets of the solar system, with Earth located a little above the others, as if it is jumping up to be called on. Finally, at the bottom, they encoded a schematic of the Arecibo radio telescope, its giant dish facing out.

In November, Drake and Sagan spooled up the giant radar and blasted this message toward the exact middle of a star cluster called M13, which has hundreds of thousands of suns inside. Its waves won't really boomerang back like the military's, and given that M13 is 22,000 light-years away, the broadcast still has a ways to go. By the time it gets there, M13 will have traveled to a different spot in space. The message won't pierce its heart. And isn't that such a human problem—a connection missed, a message lost, because the universe has moved on without us.

After its role in attempts to contact aliens, Arecibo watched for their attempts to make contact. The telescope has participated in NASA's short-lived SETI program, the University of California Berkeley's long-lived Search for Extraterrestrial Radio Emissions from Nearby Developed Intelligent Populations, and the SETI Institute's privately funded Project Phoenix.

Those never caught anything alien. But in 2001, a crop circle appeared in the UK, near the Chilbolton radio telescope. It showed the same binary message that Arecibo had sent out, returned like a radar ping. People call it the Arecibo Answer.

It wasn't, of course. Crop circles are hoaxes (why would aliens travel across the universe to crush a few agricultural products?). For instance, two British men—David Chorley and Doug

Bower—confessed in 1991 that they had made much of England's ag-art since the late 1970s, and then demonstrated their method for reporters. After that, copycat designs became more complex, with GPS and lasers potentially playing a role in modern formations.

But the Arecibo Answer, its creators never identified, holds within it a comforting idea: That they heard us. They were here. They cared enough to say, "We got the message, and we see you for who you say you are."

If a real Arecibo Answer ever did come, many people believe that the government would hide evidence of it. That occlusion is etched into popular culture. It's the premise of *Close Encounters of the Third Kind*, a movie in which the government fakes evidence of a toxic chemical spill so that they can evacuate an area for a saucer landing spot. It's the premise of the *X-Files*, a television show in which two FBI agents investigate inscrutable cases and catch clues that aliens are here and elites know. It's the basis of myths about Area 51, about Roswell.

Whether this be cultural cause or effect, many Americans don't find *some* version of those stories impossible: A 1996 Gallup poll found that 71 percent of Americans think the government is hiding information about UFOs. A June 2019 Gallup poll found that statistic holding pretty steady, at 68 percent. But, says Gallup, "it appears that only about half the number who think the government is hiding something about UFOs think it is covering up information about alien space landings, specifically." Sixty percent of people say all UFO sightings are just human activity or natural phenomena. Sixteen percent of people, though, reported having witnessed something.

Still, that means about one in every six people you meet is a UFO seer. And, according to the poll, one in three people believes UFOs are alien spacecraft. If you're not in either of those groups, you're probably scoffing right now. But if you're scoffing right now, statistically you should go ask three to six people you know. At least one of them will probably say, "Well, yeah, sure."

This widespread belief is probably why, when I used to give tours at Green Bank Observatory, where I worked for a couple of years, someone would usually ask, "Have you found aliens?" At the time, I said some version of "Of course not! You'd know." Then, I'd think to myself that if the observatory had seen *something* and hadn't disclosed it publicly before, they certainly wouldn't do so in response to a random tourist's question.

Others felt similarly frustrated by the ask. Before my time, another tour guide would leave a little green man toy on the side of the road. In the middle of the tour, he'd stop the bus in mock shock, burst through its levered door, nab this extraterrestrial, burst back inside, and proclaim, "I've found aliens!"

I'm more sympathetic to the question now. In Green Bank in particular, there was extra reason for people to associate the telescopes with alien contact: Green Bank, a tiny town of around 250, is the birthplace of SETI. In 1960, Drake—he who designed the Arecibo message—pointed an eighty-five-foot-wide telescope at two sunlike stars. Maybe somebody lived there, he thought. Maybe they were sending out messages using radio waves—a technology we'd developed earlier that century—and this radio telescope could pick those broadcasts up.

All he got was static. But since that first shot, this rural place has made more attempts to get in touch with whatever might be out there. First, Drake and other scientists held a strategizing conference at which he presented his now-famous Drake Equation and charted a course for alien searches. NASA scientists did early SETI experiments using the 140-foot telescope, and some of those same scientists continued the work through the private Project Phoenix, which the SETI Institute ran. Today, researchers with the Breakthrough Listen program—a $100 million, ten-year initiative funded by Russian billionaire Yuri Milner—continue the search. Given those decades of work, it's not crazy to think that scientists might have seen something. And it's also not crazy to think that they might have kept it a secret.

If you say that to scientists, though, they'll protest. They're bound by a culture of openness: They publish results in journals; they archive their data for all; they're part of a globally collaborative community; they want to find aliens as much or more than anyone else; they have a public protocol for what will happen if they do get a signal; *they're* not the military. And that is all true. But given the national security collaborations of the past, and given the government's ability to compel mouths shut—and its past concerns about mass hysteria—it's not 100 percent inconceivable that a hypothetical situation exists in which a scientist would stay quiet. Or, as some—including DeLonge—suggest, perhaps the government wants to slow-roll disclosure, to get people used to radical ideas before laying the full scale bare, and perhaps scientists could be convinced that was a good idea. Still, it's not likely. They don't tend to be great at following orders.

Astronomers are generally willing to talk about searching for radio broadcasts from distant alien civilizations. But their tune changes key when those broadcasts get associated with actual aliens, or with ships those bodies might pilot. Searching for electromagnetic radiation is science; searching for ships is not. Shore up those borders.

But rumors of more active, embodied aliens have always existed in Green Bank. Take a headline from the late 1980s, when a 300-foot-wide telescope collapsed onsite. It had been built in haste at the beginning of the observatory's life, meant as a stand-in, slapdash instrument to sub in while astronomers built their real, sturdier one. But it had worked so well that they'd kept it around for decades longer than intended. Then one day, in a situation we can all probably identify with, it simply fell to the ground, the victim of metal fatigue.

Or so they *said*.

Or so *they* said.

The *Weekly World News*, a tabloid, said differently.

"Space aliens zapped the enormous radio telescope at Green Bank, W. Va., with a powerful laser to keep scientists from monitoring their activities in the northern hemisphere," the article proclaimed. It attributed this hypothesis to a mysterious Swiss astronomer named Peter Voisard (of whom no record exists beyond this article). The destruction, the alleged Voisard allegedly said, "qualifie[d] as the boldest act of extraterrestrial aggression in the history of the world . . . We know that extraterrestrials have shot down planes and abducted people but this is the first time they have been brazen enough to destroy a government research facility . . . The Americans must have been learning too much about them."

Telescopes like these do have an aura of the otherworldly. Especially when they look where or at what you can't. Sunspot's solar telescopes stare at the Sun. That's a no-go for you if you like your eyes. And radio telescopes pick up on electromagnetic radiation beyond the range of our personal detector. As they pivot from this place to that, scanning across the sky, they resemble autonomous monsters, seeing secrets.

This is perhaps most acute at another radio observatory, located on the Plains of San Agustin in New Mexico. The Very Large Array, made of twenty-seven separate antennas, lies about four hours from Sunspot. It's also about four hours from Roswell, and some believe this valley is actually where the "Roswell" crash took place.

Out here, empty eeriness penetrates the landscape. Set alone in the rattlesnake-riddled sand, the telescope's antennas sit in stark relief to their surroundings. They look like they don't belong. Like they just dropped down from the sky.

In the book *Contact*, and the subsequent movie starring Jodie Foster and Matthew McConaughey, it is from here that the main character—radio astronomer Ellie Arroway—detects the *whomp* of a radio broadcast from an extraterrestrial civilization. The Pentagon, and guys with guns, join the fray shortly thereafter.

In reality, the VLA has never really done SETI work. But, from the outside, it can be difficult to understand what the telescope *does* do. Diving into the papers and data that scientists publish doesn't help: Unless you're an expert, they're impenetrable, written in the stilted, jargon-heavy vernacular that scientists use to communicate with each other. Even turned into pictures and press releases, the results sometimes remain hard to grok. In the absence of evidence we can understand, it's easy to come to alternate conclusions. And it's not a giant leap to think maybe there is, in addition, absent evidence.

IV

As I squinted into the darkness that lay between me and Sunspot, from my campsite at Cathey Vista, I had no idea what was going on over there. My mind spun through possibilities, filled in narrative blanks. Angry ex-boyfriend with a bomb threat. Angry ex-employee with a gun. Spies. A data breach. Threat from an unhinged conspiracy theorist who thought the astronomers were hiding something.

I lit the hard mesquite wood that the "if it's dark, it's dangerous" guy had sold me and perused the Internet's conspiracy forums for clues. When I went to sleep, I dreamed that strange beings, mostly human, waited just outside my tent, intending to do me harm.

In the morning, I saw that the land around me—which had felt so sinister in the dark—was actually a beautiful, welcoming place. White aspen trees and the usual range of high-altitude conifers shaded a trail lined with bright rock. Mossy groundcover bounced my shoes into my ankles as I gathered provisions.

Within a few minutes of sucking down instant coffee, I was off down the trail to Sunspot, rehearsing what I would say to whatever FBI agents met me. *I'm a reporter, I'm on National Forest land, this is public property.* I felt righteous, and self-righteous, in my search for the truth.

The trail led me along a sort of ridge toward Sacramento Peak, then thrust me into what, back East, we'd call a holler, before spitting me out onto that heliport that Sunspot's early residents had constructed. Soon, on my left, I saw the spike of the Dunn Solar Telescope, which looks like a spaghettified pyramid pointing at space. A house of worship, in its own way, it shoots 136 feet above the ground and also more than 200 feet *into* the ground. I almost couldn't believe I was so close—to the telescopes, to the mystery—and expected to be stopped at any time. But I trekked on, unmolested, until the trail ended right at the town itself.

Sunspot was a kind of place I immediately recognized: an astronomy town, just like Green Bank. The people who live here work here. The people who work here live here. The houses are plain-and-simple structures. The remnants of the site's scientific history remain scattered around, the old juxtaposed with the cutting edge.

Strangely close relationships form in places where telescopes perch atop mountains. I've never had more of a social life than I did when I lived in Green Bank, where astronomers and engineers and grounds-keepers taught and attended mountain bike trips, Zumba classes, card nights, lawn bowling tournaments, hikes, and dinners just because. Because when there's nothing going on, you make your own narrative. Years after I left Green Bank, where I only lived for two-and-a-half years, driving back into town feels like coming home. So while I can't say I felt *at home* in the abandoned Sunspot, I did feel like I *got* it.

No signs or caution tape warned me not to enter, and so—now practicing my "no posted warnings" defense—I stepped onto the grounds, onto a road named Coronal Loop, which is also and not coincidentally the name of features in the Sun's atmosphere.

I would learn later that "evacuation" may have been a generous term for what happened here: Only around ten people worked at Sunspot when the FBI showed up. The National Solar Observatory had moved its headquarters to Boulder, Colorado, shifting its money and its gaze to a new facility in Hawaii. The National Science Foundation would soon cut funding to Sunspot, and a consortium of universities would step in and take over. It's a story familiar to most astronomers. Today, scientists outsource telescopes to even more remote mountaintops and run them remotely. The money tends to go toward huge projects, leaving insufficient funds to keep up the older ones.

In the town, I walked by observatory domes and the visitors' center and an old grain silo that astronomers had repurposed into a telescope. I kept expecting someone to stop me. To show up and ask what I was doing here. To tell me I'd already seen too much and had to go. But no one did. No one even seemed to know, or care, that I was there. This lack of security deepened the mystery even more: If there was a threat, why didn't anyone want to keep me safe, or ensure I myself wasn't the danger?

The maddening thing was this: I knew the truth was out there. But whether I could access it was another question entirely.

The first true-ish answer came on September 14, from an anonymous poster on the message board 4chan. 4chan is an objectively awful place—full of racial slurs and rape threats and nationalist memes and bad plans to break the world. But also, sometimes, genuine information. And more than a week after Sunspot closed, a who-knows-who claimed to have gotten details about the evacuation from a friend at the FBI:

- NM is a hotspot for child pornography production and distribution
- the server at the observatory had been used to digitally transfer child pornography
- the FBI regularly tracks the distribution of child pornography and detected suspicious activity on the server
- the FBI investigated the facility and discovered a laptop connected to the server that was being used to download/upload child porn for 23+ hours every single day

- the laptop was determined to be the possession of a janitor who worked at the facility
- he was taken into custody
- someone made violent threats to the observatory as a result of the FBI investigation
- the observatory chose to shut itself down temporarily while the FBI investigates the seriousness of the threats. You heard it here first. Screencap this post.

Pretty much no one outside of 4chan (except a few redditors and Above Top Secret posters) noticed this declaration. But three days later, the observatory reopened. And the day after that, the courts unsealed a warrant and accompanying affidavit from a US District Court in New Mexico, claiming pretty much exactly what the anon had said: A Sunspot janitor had been downloading and distributing child pornography from the site. The janitor had exhibited threatening behavior, officials said, and as a precaution, observatory managers had evacuated the site.

In terms of headlines, this was case closed. But that didn't stop the conspiracy theorists from theorizing, from suspecting that this story wasn't *the* story. And although child porn distribution is most likely the real reason for the shutdown, if you look at the warrant and affidavit a certain way—and consider that to this day, the janitor has not been charged or arrested—it leaves plenty of room for imagining a different narrative.

The FBI agent filed the warrant the same day the observatory closed, on September 6. It gave her permission to search a Lenovo IdeaPad laptop and associated paraphernalia for evidence of illegal media distribution. And the attached affidavit tells a strange tale: In late August, Sunspot's chief observer—the telescope runner—reported that he had discovered a laptop running in an empty office, wedged between a desktop CPU and the wall. He'd seen it multiple times over a few months. "The first time he picked up the computer, he described the contents as 'not good,'" the affidavit reads. ". . . When he opened the laptop computer he saw an image of a naked adult woman, which rapidly changed to an image which the Chief Observer described as being a female child with her face covered with a mask." But "distracted by an urgent matter," he shut down the computer, and did not report the find.

The observer did not, in fact, report his finds until after law enforcement had detected suspicious activity going back months, contacted the observatory's director, and set a trap for the alleged porn distributor.

On August 21, an agent came to the observatory and seized the laptop. When the janitor returned, the affidavit says, he wanted to clean the room in which the observer had found the computer, and became concerned about missing cleaning supplies. The janitor then allegedly ranted on about people breaking in at night to steal Wi-Fi, steal laptops, and stated that he was concerned about security at the site. It was these statements—and ones the janitor allegedly made later, about a serial killer in the area—that the observatory found dangerous, threatening. That justified the evacuation.

This story, though, doesn't quite add up. First, agents had the computer beginning August 21, and the janitor allegedly grew agitated then. Why would they not have evacuated immediately? Why would he only be a threat when the FBI started to search the computer they already had in their possession—but not before? Why would the FBI need to descend upon the observatory to execute a warrant that was mostly about a computer that was in their field office? And besides, if agitated child porn suspects shut down towns, wouldn't cities be closing left and right? If none of that makes you suspicious, consider this: The listed serial number for the computer does not exist in Lenovo's database.

And then there's the matter of the chief observer—why is he not a suspect? Why did he delay the reporting process? Why would a computer involved in illegal activities not have a password? Why would its illegal material appear on-screen when the observer opened it?

Anyway, more than a year later, at the time of publication, no one's been arrested. If someone else is a suspect, who? If the evacuation was a cover, what for? What *really* happened in Sunspot?

In reality, the truth is probably *some* version of the story we heard in the news. But, regardless, there's almost certainly more to it that we don't know, and probably never will.

Someone somewhere knows, though. And that is infuriating. The curiosity gap, the space between what we do know and what we want to know, in conspiracies as in clickbait headlines, ignites our minds. Plus, there's the idea that we can build a bridge across that space. For a headline, it's easy: Read the article. For a conspiracy theory, good luck. You'll likely end up with half a bridge and a never-fading sense, standing on its unfinished edge, that there's something *more* here. Or something more out there.

I wasn't the only one to sneak into Sunspot during its closure. A man from "420 TV Freedomist Films" walked right through the front gate and onto the grounds, doing a mostly boring tour of the totally empty town.

When he approaches one of the locked lab buildings in his resulting video, he points his camera at a trash can inside the door. That, too, would be boring—except that a CD-ROM case is lodged in the can's maw. And it's the case for the 1997 game *X-Files: Unrestricted Access*.

"This is some kind of joke," he says.

Maybe it's a joke he played. Maybe the FBI, or an astronomer, played it. Or maybe it's a clue: *Unrestricted Access* was supposed to be a genre-bender, a game that was also a database of information, with ways for players to investigate cases and real footage of Mulder and Scully filmed just for this disc.

"Prepare to enter the world of the *X-Files*," said the promo video. ". . . It's going to be like being inside an episode of the show, and being a character."

This, you think, is how it begins.

TEN

ALL ALONG THE UFO WATCHTOWER

I

Despite New Mexico's ufological notoriety, it doesn't actually have the most UFO sightings in the states. That crown—per capita, and according to some calculations—goes to Colorado's San Luis Valley. When Judy Messoline bought 320 acres there in 1994, she just wanted a quiet life, a ranch far from the din of Denver. This version of her existence did not go as planned: Instead, on that land, she runs a tourist attraction called the UFO Watchtower, a platform from where people can scan the sky for evidence of a phenomenon Messoline didn't even really believe existed when she built the structure.

Since its founding in 2000, the UFO Watchtower has become many things for those many people: a pilgrimage point, a convenient pit stop, a debate floor, a community gathering place, and a haven. The watchtower draws thousands of believers, skeptics, agnostics, spiritualists, self-proclaimed psychics, and antagonists each year.

And the money it generates has also allowed Messoline to hold on to her land through lean times.

If you've never been to the UFO Watchtower, you can be forgiven: It's in the middle of nowhere, in a part of the country far from most other parts of the country. If you're coming from the north, take Highway 285 and slide down over Poncha Pass into the San Luis Valley, curving eventually onto Highway 17. To your left, the Sangre de Cristo mountain range rises more than 14,000 feet into the sky, its Teton-esque peaks staying snow-capped well into the summer months. To your right swings the valley's vast sprawl: It's 60 miles wide in most spots, the view to the west cut off by the distant San Juan mountains. In this 100-mile-long, vaguely pentagonal pancake, vegetation rarely rises more than a foot or two high. You can see seemingly forever. It's the largest alpine valley in North America, and despite its desertiness—the sandy ground, the crackly plants, the cacti—a giant aquifer lives beneath it. The Rio Grande, born in the mountains just above, sluices over its southwestern end, and 20,000 sandhill cranes sojourn there seasonally.

Perhaps the valley's strangest feature, though, are the Great Sand Dunes. From a distance, they look like someone took a peach paintbrush to part of the landscape. But once you get closer, they reveal themselves for what they are: a slice of the Sahara stuck right up against the Sangre de Cristo mountains. Long ago, the valley was filled with lake water. Once the water disappeared, only the sand was left, and the prevailing (unrelenting) wind pushed it up against the range's lowest passes, where storm winds—which blow the other direction—whistle through. The sand stays trapped, building itself up vertically. Colliding winds and colliding time periods create a land of contradictions.

But don't get ahead of yourself. Once you descend toward the valley, you have to drive another 50 miles before you hit the entrance road to the UFO Watchtower. Warning you of the impending attraction, a lime-colored alien stands guard. Pegged to a wagon wheel, it points down the dirt road. Big beams, like those fronting most ranches, form a kind of portal, where a handwritten sign beckons. "COME ON IN," it says. "Explore the UFO Phenomenon at the UFO Tower." If you come from the other direction, a larger, yellow sign shows a cowboy alien riding some kind of warped saucer, hat thrown back like a rodeo pro. "Ride the Cosmic HWY to the UFO Watchtower," this one says. If you oblige these invitations, you'll travel briefly down a dirt road and turn into a campground. Other aliens stuck to other wagon wheels point to the watchtower itself.

The watchtower doesn't amount to much, really: It's just a tenish-foot-high metal-grate platform, scaffolded with metal poles, shaped like a squared-off U. In the U's gap, a saucery dome rises to about the height of the watchtower's walls. This dome is the visitors' center and gift shop, which Messoline has mostly run on her own since 2000.

On a Saturday in June 2019, the UFO Watchtower is busier than I've ever seen it. Five or six carloads of people wander the grounds, new Subarus and CRVs pulling in to replace them once they leave. Summer is the busy time here (winter, with its average low of around 2 degrees Fahrenheit, is brutal). Many visitors have come or are on their way to Great Sand Dunes National Park, a quick drive down the road.

This weekend, Messoline sits quietly on the concrete porch of the dome, watching people take in the place she has made her work and her home for the past two decades. Every so often, she turns her head toward the dome, as the woman running the register—Annie—picks up Messoline's book and gestures toward the founder. "She's the coolest lady in the world," says Annie. The book is titled *That Crazy Lady down the Road*.

Messoline's journey toward building the UFO Watchtower, which she details in her book, started with a single step: into the National Western Stock Show in Denver in the nineties. Messoline had recently gotten divorced, and her roofer—Stan, soon to be her partner—suggested they go distract themselves by looking at the animals. But, just as you end up with ice cream and Cheetos if you go to the grocery store hungry, they came home with six cows between them.

The two moved the cattle to Messoline's few acres near the town of Golden, a Denver suburb that retains some of its Old Westiness and houses the Coors brewing facility, set up against the Rocky Mountain foothills. They liked their life out there fine enough, but having such close neighbors felt tiresome. Once, a bull got out and rubbed its body against a neighbor's tree, damaging it. Humanity was causing problems, and Messoline wanted out. That's when she started watching the newspaper ads, waiting for land big and cheap enough to live on. God, she believed, would send her where she was meant to live. And that happened one day in 1994, when she saw 160 acres for sale in the San Luis Valley. She and Stan soon got in the truck and headed south to check it out.

As soon as they crested Poncha Pass, and Messoline saw the geography roll out like the promised land before her, she knew: This was the place. This was, in fact, a whole other world. It looks not just different from Denver but different from most of the rest of Earth. When the realtor got them a deal on double the acreage, they signed right away.

Things were soon to get more complicated. In western Colorado, the San Luis Valley's parcels often go for less than other ranchable areas. And that's not without reason. It's not close to much of anything, the winter weather is hostile, there's not much infrastructure, there aren't many jobs, there's not much money, there isn't much rain, and water rights are complicated.

That trouble, though, still lay ahead of the couple. Right then, the bigger problem was that the house in Golden didn't sell for a year. Once, needing to make her valley mortgage payment but only having $40 to her name (and the land not being *that* cheap), she took the money to Black Hawk, a town lit with casinos. Playing just the dollar machine, she nabbed the jackpot: $2,000. Enough.

It took Messoline years to build the house and the fence that enclose her life and the property today. But not long after she settled in, people started telling her about the UFOs. First, a friend pointed her to a book called *The Mysterious Valley* by Christopher O'Brien. For the book and two that followed, O'Brien investigated more than a thousand reports of paranormal activity over a ten-year period in the valley. It was a hotspot, and had attracted a little infamy.

But Messoline had moved down here to be a rancher, and like a practical person, she mostly put thoughts of the strange skycraft away. Still, more people started to talk about them. A woman at the pool where Messoline worked part-time. The UPS guy. Farmers. She heard about the valley's cattle mutilations, and mutilations of other animals—like Snippy the horse, found in 1967 with strangely surgical cuts and no blood on the scene, one of the first spooky killings in the country.

As people spun these stories to Messoline, she floated the idea of a watchtower. If everyone was seeing so many "UFOs," why not give them a place to keep an eye out? It was a joke at first, but as people confessed more and as the valley's notoriety grew, the concept gained gravity in her mind. She saw a financial opportunity here. And despite her doubts, others' experiences started to seep into her consciousness like drops of rain meandering down into the aquifer. One day when she and Stan were driving, she saw a white light in the quintessential UFO shape. She screamed at Stan. Stan, meanwhile, noted that it was just, in fact, the moon, rising up into the sky.

Still, even with her mind primed, it wasn't time to reap the benefits of a watchtower yet. Other events had to lead her life in this direction. Messoline lost a horse whose intestines wrapped over his spleen; a dog had a heart attack; a neighbor's dog killed some more of her cows; Messoline accidentally ran over another of her dogs; two bulls fought to the death; two cows ate innard-complicating brush. And, finally, after four-and-a-half years of running a ranch, Messoline no longer had enough money to buy feed for the cattle. The water she'd been promised for irrigation didn't exist: If it didn't rain, food didn't grow from the ground. And it didn't rain. If she didn't sell her beloved cattle soon, they would only have sand for supper.

She cried when she let them go.

II

At that low point, a friend suggested that maybe Messoline should actually build that watchtower thing she'd been joking about. And so it was that that same day, Messoline drove to the Saguache County Courthouse for the paperwork. The morning after that, she drove it and the $300 fee back. A few weeks later, she stood before the county's planning commission to pitch the idea. One of the commissioners visited the site, watching while Messoline explained her vision, sculpting it from the air. There would be a gift shop, a deck, camping spots, signs out front to grab tourists' attention, admission fees, T-shirt money. Business income and heads in beds for an area short on all of that. When Messoline met again with the commissioners, they told her to go for it.

Behind her in the courtroom, though, sat a reporter, taking notes on this watchtower business. That reporter told another reporter, who wrote a story that got picked up by the Associated Press, which went what we would now call viral. Messoline did interview upon interview. And like magic, the UFO Watchtower became a thing before it even was a thing, fervor whipped up out of mere words. Given the press, Messoline expected that lots of people would show up to the watchtower's opening. Instead, on Memorial Day 2000, just some UFO buffs and a few curious onlookers wandered under the dome, and stood on the deck that in its current form can hold about fifty people if they are willing to be friends.

A month in, bills were coming due and not enough people were coming in. Frustrated, scared, Messoline yelled at the sky. "You guys, this was *your* big idea," she wrote of the experience. "I need at least $100 a day in sales to make this work!"

She got it. And soon, things started to look up: More and more reporters came through—the number of reporters still disproportionate to the number of actual visitors. But perhaps because of that coverage, visitors soon did come, and they started leaving more money (she got her first $100-bill donation in 2003).

The watchtower has since had its share of difficulties: nudists sneaking in after-hours, a metaphysical music festival that was full of drugs (duh), a theft of the alien out front ("ALIEN ABDUCTED" proclaimed the newspaper). But over the past twenty years, visitors have claimed to see hundreds of UFOs from the tower, their accounts gathered into an overstuffed three-ring binder.

Perhaps the most intriguing aspect of the tower, though, is not what's above but what's below: a rock garden littered with meaningful detritus. Psychics in the tower's early years told Messoline that two energy vortices swirled right next to the tower, entrances to parallel universes. If she wanted to help people absorb their energy, she could use rocks to outline them and set up benches for people to sit, feel, meditate, contemplate. Today alien sentinels stand guard around the garden. There's a silly purple one with eyes literally as big as its head, a sasquatch, E.T. from *E.T.*, a Grey wearing sunglasses, two lithe beings—one green, one black—set on concrete platforms and decked out with accessories. They, the explanatory sign suggests, represent the two beings who guard the vortices.

This same sign also requests that visitors leave something of their own, depositing their energy to swirl with the universe's. And so it is that years of junk lay scattered around the supposed portals. The whole area is kind of like a backward pointillist painting—nonsensical and chaotic from far away, deeply human and ordered close up.

People have left at least two copies of the movie *Homeward Bound*, in which sentient creatures (two telepathic dogs and a telepathic cat) find their way back to their owners, trekking across the Sierra Nevada mountains to San Francisco, following either some natural sense of direction or the odor of love and abandonment. There's a Book of Mormon, a Bible. There are sunglasses, toothbrushes, batteries, beer openers, coins, sunscreen bottles, bungee cords, bingo dotters, a fishing bobber, toy cars, a ski, a Jay-Z album case, dolls that look like Sid from *Toy Story* got to them, airplane-sized liquor bottles, jewelry.

Anyway, it looks like trash. Almost nothing in this garden has much extrinsic worth. But each item meant something to the person who left it here. In that way, it is like a UFO sighting or a dream: meaningful mostly to the person it belongs to.

The leaving of an object also imbues it with personal worth it may not have had before. The person who left the toothbrush that had been living in his glovebox for five years wouldn't have remembered its existence if it hadn't disappeared from his life. That Smirnoff bottle? That hand sanitizer? The garden is a long list of disposables, meant to be tossed off and forgotten. But now the leavers will likely recall that bottle, that brush, and know it's still out there somewhere. *It's not that they appear. It's that they disappear.*

III

She's the coolest lady in the world," I hear Annie say again as I walk inside the dome. She's telling a couple about the twentieth-anniversary party the Watchtower will hold next year, a weekend-long celebration for which Messoline so far refuses to charge an entrance fee. Messoline still sits just outside the door, staring distant-eyed into the garden, occasionally getting up and hovering in the threshold. The two women have an interesting dynamic. For example:

Annie drops something and gives Messoline permission to fire her.

"Oh, Annie," she says. "I'm not gonna fire you."

Annie then periodically apologizes for other small things, like forgetting to request that someone sign the guestbook. "It's *OK*, Annie," Messoline emphasizes, every time. It seems like a conversation they have on recursive loop—one that is, like any habit, comforting as well as tiresome.

Messoline eventually wanders back toward the house, a campground and the width of a dirt road away. Annie is alone.

"May 28 of last year, Judy fell and shattered her femur," Annie now says conspiratorially. Messoline had been pulling down hay for the cows and horses, not realizing some twine had wrapped itself around her ankle. When she pulled the hay, it pulled back on her, tossing her to the ground and fracturing her leg bone in seven places. At that point, Annie, who had been cleaning the bed and breakfast on the property for three years, got a frantic nighttime phone call.

"Can you come help?" asked a friend of Messoline's.

"Of course," said Annie.

Back then, she was just helping Messoline with household things, because if Messoline saw dust, she would clean dust, even if it killed her. Annie called this time "Judy-sitting."

As she gives me this vocab lesson, a family arrives at the watchtower from Canon City, which is about 60 miles northeast, at least if you could warp drive over the Sangres and the Wet Mountains. Annie tells the two shy kids that they can each pick out a tiny alien figurine for free. "We screen-print all the T-shirts ourselves," she says to the parents. "Except for one."

"Which one?" the matriarch asks.

"The one that's really cool," Annie says. And, indeed, there's a marbled purple garment, a sentient Grey detailed like a computer-generated portrait, looking out from the torso: the alien inside you. Messoline saw them in Texas, bought three stores' worth, and drove them back up to Colorado. After she encourages the family to sign the guestbook, they wander out into the garden.

"How does all the stuff out there not blow away?" I ask, knowing that the winds in the Mysterious Valley sometimes merit the term "hurricane-force" (as Annie, who's from the Windy City, put it, "Chicago ain't got nothing on this").

"Magic," she says, smiling. "I have seen dust devils come through and pick things up and set them back down in the garden."

As the visitors wander among the littered talismans, Annie continues her story. Messoline has mostly run the watchtower alone for the past nineteen years. "Now, I told Judy," she says, "'You're seventy-four, you're retired, you can have two days a week.' Annie would work the other five. And if Messoline passes away—which she couldn't possibly, says Annie, seeing as how she's immortal—Annie would find a way to keep the tower going.

"This can't end with Judy," she says. It means a lot to people, to her.

It's clear she means this, and not just because the watchtower gives good, steady work—a paycheck in a place without many jobs.

"How'd you come to be here at all?" I ask.

Annie pauses, fiddles with the flimflam beneath the counter, and invites me to grab the stool that's sitting up against the door.

"The universe brought me here," she says, finally, without the slightest brush of irony.

Annie got pregnant when she was fifteen, then became homeless for a while. She gave birth to her first child and moved from Chicago to Missouri as soon as she could, at age seventeen. "I did not want my kids growing up the way I grew up," she says, shaking her long red dreadlocks.

She and her kid lived in a place called Nevada and pronounced "Nevayda," then in Branson, the Vegas of the Ozarks. With its dinner theaters and its Ripley's Believe-It-Or-Not realness right next to beautiful nature, it's hard to distinguish the for-show and the for-real. But they liked it. She got older; she had two more kids; they settled in for the long haul.

"Then I made every bad decision I could make," says Annie. It doesn't matter what kinds of decisions. What matters is the outcome: "Two years of chaos," she says. Those two years ended with her and her husband—kids still in Missouri—moving to Barstow, California, a Mojave-Desert town where a friend promised work that didn't actually exist.

"We ended up pushing a shopping cart," says Annie.

They lived in a tent behind the Denny's, taking shelter near America's diner in this superheated nonmetropolis at the intersection of I-15 and I-40. "I was terrified to go to a shelter," says Annie. "It's gotta be like on TV with crazy people and children screaming, crying."

But her shoes had holes in the bottom that she'd had to patch with her husband's tube socks. The ground—even the grass—was so hot it burned.

"The universe finally said, 'All right, get in the backseat: I'm driving,'" says Annie.

They gave in and went to a shelter, which was nothing like what she'd feared. Soon, she was hopping on a computer. The shelter director would pay to send the couple to stay with friends or family if they could find some. Not knowing who to contact, Annie got on Facebook and put it all out there. "This is what I've decided to do; I've made every wrong decision; I don't know what direction to go; this is where I'm at; this is why I'm where I'm at; any input would help," she recalls. "I don't know what to do."

Soon, friends in Alamosa—23 miles from the UFO Watchtower—said the couple could stay with them. There was work, they promised; it was beautiful. The shelter director bought their bus tickets, and they arrived in the nearby town of Walsenburg at three o'clock in the morning.

Walsenburg was once known as "the city built on coal." Also: the backs of laborers. Workers—forced to do ten-hour days without proper ventilation for explosive gases and only paid for the full loads of coal they surfaced—staged a series of strikes in the early twentieth century. One, at a tent city south of Walsenburg proper, resulted in the deaths of twenty-one miners.

Today, it's still an Old-West-looking town, with a one- and two-story southwestern main street, the muted pastel stone storefronts sheltering antiques and lumber and espresso. They arrived the last day of August, just in time for the town's car show. That first morning, she and her husband walked around, admiring the chrome and glossy paint of cared-for, well-kept vehicles. The cars glinted in the sun, their owners' lives so alien to Annie's own.

But after a while and much frustration, things started to look up: Annie soon got a job housekeeping at the Holiday Inn. The couple rented their own place at Splashland, a natural hot springs corralled into a concrete pool with a corkscrew slide. When Annie went to sleep, she layered her clothes to make up for the cold and lack of bed. She started cleaning Messoline's bed and breakfast. She and her husband were doing well. They had stored up some money and were about to use their stash to move, when, on Christmas night, she heard her sixteen-year-old son had been hit by a pickup truck. Getting back to him (he ended up OK) drained their savings. She was stuck in Colorado, something bigger keeping her here.

Not long after, the well at Annie's rental froze up. Messoline offered to rent her one of the houses on the Watchtower property.

"I can't afford it," Annie told her.

"What if I made it affordable?" Messoline asked.

It was, for the first time in a long time, a safe place to land.

IV

As Annie tells her story, the wind comes up like the bong of a clock: seemingly all of a sudden, though predictable in its arrival if you'd been paying attention. It whips up particles from the dunes, grains of sand shooting into the air and looking, from this distant post, like a goldenrod fog.

At the Sand Dunes' visitor center, you can see a map of the biggest sand hills, up to 750 feet tall. Their names and mapped shapes lend them an illusory permanence. The truth is, they constantly shift and overwrite themselves. They get abraded, abrade each other, abrade tourists. They're basically like us: beautiful, horrible, strange, offensive, and always changing. The park map will be irrelevant to far-distant future generations. But that's true of anything, even the most solid-looking stuff: Time will erode and accrete, making everything different and then gone.

A group enters the watchtower dome to seek shelter from this time-shifting wind. "Do you want to see the alien body?" Annie asks them as she reaches beneath the glass case and reveals a take-out container with a plastic lid. A mummified creature rests on a paper towel inside, its head twice as big as seems right and twisted unnaturally backward. Its four legs curl fetally against itself. "Alien body," proclaims Sharpie from a piece of masking tape.

"Do you know what it is?" Annie asks.

"A goat?" says one visitor.

"Haha, yeah," says Annie. "It's a goat."

When the guests go back to the garden, Annie continues, telling me about Messoline's forthcoming second book and its big reveal about a ravine out back.

"Judy's had several psychics tell her that's the site of an ancient ship crash," she says. When Messoline asked them how big it was, one responded, "How much land do you own?" It is, they tell her, a mile long, bobbing up and down beneath the Earth.

"Makes total sense," says Annie. "Underneath us is a giant aquifer."

She pulls out drawings the psychics have made, which show a fairly classical flying saucer. (It is only then that I realize she is not talking about a boat.)

This tale is interrupted by Annie's youngest son, now eighteen, who lives with her now and works on Messoline's ranch. He stands in the doorway, arms tan and strong from outdoor labor. He's hot-eared about an insult a coworker made. Annie calms him, dispensing very true wisdom about how some people are just dumb. Eventually, he asks her if he can take the car to the gas station a few miles up the road, apologizes for interrupting, and says it was nice to meet me.

He seems like a nice kid who's had an already complicated life, now watched over by the watchtower.

The sun, by now, has slid to the western side of the sky, closer to the San Juans than the Sangres and the Sand Dunes. And as Annie counts cash and sorts out who's filed a campsite form, we finally talk about UFOs, a topic that always seems to save itself for last.

Sure, she's seen stuff, she says. A black helicopter disappeared into a gap in the mountains, maybe into the secret underground base some believe exists beneath the 14,344-foot-tall Mount Blanca. And then there was the time, last summer, when she was sitting on the patio of her safe house by herself. Seven different wildfires burned around her, including one that grew to 108,045 acres in size. The wind, for once, wasn't. "All the smoke was like smokestacks," she says. "I'm sitting there smoking a cigarette, watching the world burn around me, thinking how amazingly beautiful and horrible it was at the same time."

A cigar-shaped ship then came out of nowhere and floated—didn't fly—for fifteen minutes, until Annie's phone rang. And then—blip—it was gone.

Annie continues counting campers' cash, but I can't stop thinking about the world burning, literally and figuratively. It often feels lately like it *is*—both ways. Actual fires are bigger and faster and meaner and more unpredictable. Climate change is altering human habitation forever. Political stability is elusive. Everyone is mad at each other—even if not this moment, definitely after they get on Twitter or Facebook tomorrow morning. Mass public shootings happen around twice a month. Companies harvest data, insinuate themselves into our homes and relationships. Ships blow up; drones go down; the threat of nuclear annihilation colors the world like it hasn't since the Cold War.

The Bulletin of the Atomic Scientists keeps an estimate of how long we have until civilization is over. It's called the Doomsday

Clock, and it's been held at two minutes to midnight since 2018. The timekeepers call this the "new abnormal," explaining:

> Humanity now faces two simultaneous existential threats, either of which would be cause for extreme concern and immediate attention. These major threats—nuclear weapons and climate change—were exacerbated this past year by the increased use of information warfare to undermine democracy around the world, amplifying risk from these and other threats and putting the future of civilization in extraordinary danger.

Doomsday, at least in this admittedly unofficial quantification, hasn't been this close since 1953—the year the Soviet Union tested its first hydrogen bomb.

Everything feels precarious. Society seems to teeter on the sawtooth ridges that connect the Sangres' peaks. It's not hard to imagine—even if you're not a conspiracy theorist—that you're living at the end of days, or at least on the precipice of some large cultural change that doesn't feel like progress.

But as much as a fire destroys, it also creates. It clears out old debris so new stuff can take root. It wipes out plants that sucked up too much water. Some trees even need it to spark their next generation. And on the small scale, fire means safety, coziness, company, nourishment. It means you are not alone. Seen from a distance, fire lights the way home, flickering in and out of existence on the horizon.

V

The proverbial fires of places like the San Luis Valley draw people—people like Annie, like Arnu the Area 51 investigator, like Dalton the A'Le'Inn bartender. There's something magnetic about them. If you ask believers, they will sometimes call these places "hot spots," areas of Earth where paranormal activity comes in concentrate. According to O'Brien, the *Mysterious Valley* author, hot spots tend to share a few intrinsic characteristics: weird geophysics, weird weather, designation as sacred to American Indians, modern subcultures around the paranormal and occult, proximity to military facilities, and a population immune to UFOs because they see them so often.

The San Luis Valley is a prime example of such a spot. It's a galaxy-wide space with mountains that rise ragged from nowhere and dunes that belong in a different landscape. The nearby town of Crestone is a spiritual center, having accreted creeds of all sorts as time has gone on and its reputation has spread. Military officials have long tussled with residents, wanting to overfly lower and faster. It's cold, windy, wild. Tewa Pueblo stories say Earth's earliest humans arrived through a hole in a lake close to the Sand Dunes. For the Navajo, the sacred Mount Blanca delineates the eastern end of their world.

Other hot spots have similar stories. Northern California's Mount Shasta, a not definitely inactive volcano, rises solo 14,179 above sea level, from a low valley. You can go on vortex-hunting tours. Some people say Bigfoot lives here; others say an alien race called the Lemurians does. Nearby, the Oregon Air National Guard's 173rd Fighter Wing trains F-15 pilots. It's cold, windy, wild.

Devil's Tower, the place where the aliens land in *Close Encounters*, rises in a monolith nearly 1,300 feet tall. Its parallel cracks, running its nearly quarter-mile height, look as if a giant monster scraped its fingernails through the rock while it was still wet clay. Ellsworth Air Force Base is about 90 miles away. It's windy, wild. Devil's Tower and Mount Shasta are both sites sacred to American Indians, and have been since long before any white people came and instituted silent retreats.

Most of all, these places are geologically strange. Otherworldly. They *all* look like they don't belong. They are beautiful, horrible.

Beckoning, hostile. They rise from nothing up into the sky. Inanimate gods.

As with UFOs, we ascribe to their strangeness a twist of the alien. They seem so weird they can't possibly just come from this world.

Maybe our own world, though, is just stranger than we are ready to believe.

After Annie starts closing up shop, I say goodbye and go to my campsite. I make a fire. I open up a can of Chef Boyardee and a book as the sun sets. It's a beach read, a sci-fi thriller about a woman who invents a device that is supposed to aid memory. In reality, it flings the users back in time. Thus chronologically deposited, they can change their own histories: undo their wrongs, make their rights better, succeed where they failed, unkill the dead, try a different spouse, make everyone—especially themselves—happier.

It sounds good at first, but of course (sci-fi being sci-fi and humans being humans) the engineering effort has unintended consequences that end with a burning world. I'm reading one of many parts in which a person attempts to live a better life by undoing their past, thus unintentionally ruining the lives of others, when I look up at the Western horizon, at the San Juan mountains dozens of miles distant. The sun streams through the clouds in a way I don't quite understand. Its beams split into seven wide spotlights, yellow and orange against the gunmetal clouds.

The couple at the campsite to my right, who have never camped before, start snapping pictures. The couple to my left, who have brought a bubble machine, get up from the red table they also brought and stare at the sky, silent. We're all facing the same direction, our pupils tethered to the firmament. Each of us is having our own version of the exact same experience. The grind of a generator, the only sound, winds through the windy air. It's one of those small, perfect moments when seconds seem not to pass, and the planet feels the most familiar kind of strange, the strangest kind of familiar.

Then, of course, the moment is gone. The sun has set on it.

None of us sees a UFO that night. None of us really came here to.

ELEVEN

IT WAS ALWAYS YOU

I

Most travelers to the UFO Watchtower have only close encounters of the first and second kind: ones in which a human sees a flying something, and ones in which that something leaves a physical impression on the world. Here, UFOs act alone, their appearance eliding the hypothetical extraterrestrial biological entities that may or may not find themselves inside.

Greater degrees of alleged contact do, of course, exist. During close encounters of the third kind, a phrase made famous by the movie of the same name, witnesses see an entity. In close encounters of the fourth kind, meanwhile, aliens abduct a human person. Ufologist Steven Greer has taken things a step further, creating the fifth type of alien experience. "A Close Encounter of the Fifth Kind is when humans deliberately take the initiative to interact with or communicate with, no matter how simply, with these life forms and their spacecraft," Greer said in a 1995 presentation. ". . . It is a new paradigm of interaction that goes from the usual passive modality or retrospective modality to one of real time, pro-active communication." He means that you can perhaps compel ET to come to you.

To discuss and study this sort of contact, Greer has created an initiative, group retreat packages, an app, and documentaries. He has even developed the CE-5 research protocols for initiating such communication with the alleged others. In Greer's view (as in that of others, like Griboski), contact happens not just because you happen to be in the right place at the right time but because your brain connects with the minds of the extraterrestrials. Close encounters of the fifth kind, as such, involve much meditation and similar intentionality to that of a vision board, inviting and directing other life forms and their spacecraft into your field of view (Greer also uses light and sound to try to "vector" vehicles toward a given area). In the fifth-kinder's view, the reality of the universe is woven with our own circuitry.

That's a lot to buy. And while Greer has a loyal following (plenty of people pay thousands of dollars to go on bilateral expeditions with him), if you Google "Steven Greer is," you'll get a lot of results from more dubious people. Many skeptically programmed ufologists steer clear of Greer.

Likewise, radio astronomer Gerrit Verschuur isn't—never was—one of Greer's adherents. Verschuur doesn't think he's made alien contact; he doesn't even think UFOs are real. But he has, through an evolution in his scientific studies, come to similar conclusions about consciousness and the universe. To get there, he had to veer from traditional science, fly in the face of its culture, and let go of the things he thought he knew.

The sky had no moon the night Verschuur drove an antique Checker cab toward a 140-foot-wide radio telescope in Green Bank, West Virginia. The car ran on diesel and had no electronics so that spark plugs and onboard fanciness wouldn't spike a signal into the sensitive astronomical instruments. When Verschuur pulled up to the control room, the only sound in the whole Appalachian wilderness came from the motors inside the telescope. The hum permeated the spacetime around him. In his book *Is Anyone Out There?*, Verschuur describes this experience and the much stranger ones that followed.

With a beard and a floppy haircut, he could have been a nineties skateboarder if it weren't 1972 and if he weren't a well-known radio astronomer—a sort of pioneer, actually. He'd made the first-ever measurement of the strength of the magnetic fields that arc among stars, and, there, he'd also discovered the coldest clouds of hydrogen gas. Though hydrogen is the most abundant substance in the universe (aside from dark matter), its radio emission stays hidden from human eyes.

During this evening's observing run, Verschuur wouldn't be investigating any of that. He had come to Green Bank to search for broadcasts from aliens. His experiment was part of a relatively new enterprise called SETI, the search for extraterrestrial intelligence, whose first experiment had taken place right here twelve years before.

He recalled the nights he'd spent at the Jodrell Bank Observatory in England, the 250-foot-wide dish of its telescope pointed up at more traditional celestial targets. As the telescope observed that stuff, Verschuur had thought about how, in the course of this work, he—right place, right time—might *accidentally* pick up an alien broadcast. Just in case, he would sometimes plug a speaker to the telescope's receiver, so he could hear the radio waves that splashed into the telescope's dish. All that ever came through was the staticky hum of the nonbiological universe, punctured by the occasional radar staccato from a nearby airport. No call from any space brothers.

Maybe tonight, he thought, walking into the Green Bank control room. "I pushed open the glass door that punctured the thick concrete wall and noticed the sign that spelled out 'Fallout Shelter,'" he wrote. "If it should ever come time to hide from man's atrocities to man, this artificial structure was about as solid as any you could find."

Inside that structure, he surveyed the racks of electronics, the tape recorder that stored readings from the telescope, and the operator in his chair. They greeted each other, and after hooking up a speaker, just like in England, the two men prepared to hunt.

Soon after the observation session started, "a squawking noise came from the loudspeaker," Verschuur wrote.

The sound shot through his spine. And then, just like that, it disappeared.

Neither he nor the operator had ever witnessed a signal like that before. Surely, it was interference from human technology—that happens with things like a plane, a satellite, a wonky electric fence. The radio waves were too strong to have crossed the vast vacuum of space, unless the aliens' transmitter was so powerful we could hardly imagine it.

"If that was ET, we'd be in trouble!" Verschuur said to the operator, joking.

Still, shaken up, he went outside to take a break and stare at the sky.

"A meteor flashed and I made a wish," Verschuur wrote. "I wanted to learn how to make contact with extraterrestrials."

He was already beginning to sense that this—a radio telescope, a speaker, a hope—was not the way.

A second meteor scraped across the sky, burning through the atmosphere of this, the only known planet with life.

II

Verschuur's attempts to travel in a new direction would later send him inside sensory deprivation tanks, toward voices in his head, into books about communicating with dolphins, onto beaches with humans who claimed aliens inhabited their bodies, and—of course—onto the shit-lists of some mainstream SETI astronomers.

Before all that, though, Verschuur was the kind of scientist who would have shaken his head at Verschuur. Around the time of his first SETI experiment, NASA sponsored a study called the Cyclops Report. It was not about a one-eyed monster, exactly, but about a hypothetical giant radio telescope, designed specifically to find alien technology. Made of a thousand 100-meter antennas working together (in an eye-like circle), it could detect the same kinds of alien broadcasts that Verschuur had been trying to tune in to. The report laid out the instrumentation, motivation, and money required to mount this kind of alien hunt. More interesting to Verschuur, though, were the scientific reasons the report gave for doing so. Such a search would advance science, sure. But also, the document speculated, we might learn how to preserve the species, lead richer lives, and act more united. Those conclusions formed the foundation of what Verschuur calls the "Salvation School of SETI."

You can still hear that perspective from SETI scientists today: If we found a much older civilization, we would infer that because *they* survived a "technological adolescence"—with nuclear bombs and/or weapons we can't even dream of—earthlings could too. And finding another literally alien species would render intrahuman differences so laughable that we'd knock it off with the wars, racism, sexism, etc. For these peace-inducing reasons, they reasoned, money sent to SETI was well spent.

Verschuur didn't buy that or the scientists' optimism about SETI's prospects for success. Wanting that latter qualm on the record, Verschuur traveled to a 1976 radio-astronomy meeting in Amherst, Massachusetts. There, he presented a paper called "The Harsh Realities of Drake's Equation." The Drake Equation estimates how many alien civilizations might exist in this galaxy that could communicate in a way that our technology could detect.

Verschuur estimated that the number was essentially zero. "That's why we haven't seen anything out there," he told me. "Because there aren't civilizations like ours."

A belief system—not scientific logic—propels current SETI research, Verschuur believes. To hope that you could find extraterrestrials inherently involves faith. The way he interpreted the scientists' reasoning, faith that the search *could* succeed, without evidence that that was likely, was fine as a motivation because contact would improve this world. The rhetoric rang like that of religion. "You think something out there is going to help you solve your problems down here," says Verschuur. ". . . You're searching for something greater than yourself and believe in that."

If you buy that, SETI philosophy and UFO belief represent two ways to have faith outside a traditional religion. In a broad sense, so do all conspiracy theories, a worldview in which invisible and powerful agents control human reality. As Karl Popper, a philosopher of science, once said, "The conspiracy theory of society . . . comes from abandoning God and then asking: What is in his place?"

Today, when "no religion" is the largest religion in the United States, with 23 percent of the population identifying that way, our collective urge to put something in God's place is stronger than ever.

Verschuur, continuing his evangelism about the evangelism of SETI, planned to attend a scientific meeting in Pasadena in 1977. There, he would present a list of alternative ways to contact aliens, beyond scanning with expensive radio telescopes. All possibilities, he thought, had a roughly equal chance of success: just about 0 percent.

He'd lived in Boulder, Colorado—a place full of New Age-y philosophers and out-there humans who walk the planet sans shoes. He'd met lots of residents who believed in aliens. He took notes on their claims, intending to present these as his alternative—satirical—approaches to SETI. "The funniest alternative was to search within oneself," Verschuur wrote. "I would be willing to accept a grant and live in the mountains of Colorado and think about [extraterrestrial intelligence] and determine if there was evidence they were communicating with me." People could also study UFOs; they could try automatic writing and see if alien voices sprang from their hands. All of these options cost essentially nothing, and they'd get the same (null) result as scientists would with million- or billion-dollar telescope projects. It was, you see, a joke.

When he presented the ideas to the conference, some in the audience laughed. And having made his point, Verschuur didn't realize this was just the start of his research.

III

In the late 1970s and to Verschuur's chagrin, NASA started to ramp up SETI projects, despite some vehement political opposition. Around this time, Verschuur walked into a Boulder bookstore. The owner knew of Verschuur's interest in alien pursuits, so he recommended a book called *The Center of the Cyclone*, by a scientist named John Lilly.

Lilly had been part of the SETI community since its beginning. He had even attended the first official meeting, in Green Bank in 1961, at which Drake presented his famous equation. Lilly was obsessed with the idea of communicating with dolphins (because of Lilly's research, attendees at the first SETI meeting called themselves "the Order of the Dolphin"). If we could talk to this terrestrial species, he thought, we would better understand how to talk to extraterrestrials. Also, some thinking goes, dolphins' very existence meant that smarts could evolve more than once on the same planet, and so intelligence perhaps came part and parcel with the evolutionary package.

Verschuur now believes a slightly different version of that. "The only thing we're certain of, biologically speaking, is evolution happens," he says. "But of what? What is evolving?"

He pauses on our phone call, as if I might know the answer or perhaps refute his assertion, which I am pretty sure biologists would.

A little like a frustrated teacher who hoped his students would have learned more by now, he says, "It's consciousness. You can look at evolution as being a steady increase in consciousness." (I am pretty sure biologists would have something to say about this assertion, too.)

Given this opinion, though, it seems strange that Verschuur feels so sure traditional SETI will fail. After all, if biology happens, and evolution happens, and consciousness drives evolution, wouldn't smart aliens saturate the cosmos?

Maybe. "But they're only going to be capable of communicating with a species at the same state of evolution as they are," he says, unless they dumb themselves down specifically to talk to us. And the chances that we're in the same state? Just about 0 percent.

This line of logic reminds me of a conversation I had with my grandfather when I was around eight years old. I already loved aliens, and my grandpa loved to red-team me about it.

"If they're out there," he said, leaning forward in his plush recliner, "why haven't they tried to talk to us?"

I had not pondered this question enough to have a ready answer. But before I even became aware that I had a thought at all, out of my mouth came the words, "They're waiting for us to get to their level."

It seemed very smart to me at the time. But Verschuur would probably call it dumb: If the aliens got a head start, they'd always be ahead, just like you'll never be the same age as someone born before you.

"We'll never catch up," he says.

Verschuur decided to take the plunge and make the purchase. Soon, he was reading about Lilly's experiences. Which, mostly, involved LSD, particularly taking LSD in isolation tanks—devices that he himself had invented in 1954. These vessels, filled with shallow, salty water, redact the sound and light of the outside world. In the saline liquid, a human is perfectly buoyant. A human hears nothing. A human sees nothing. A human feels nothing. A human reaches perfect equilibrium with its environment.

Lilly hoped that by slicing off contact with the outside world, he could reach deep inside himself, really get to know his own consciousness and also his unconsciousness. A hit of acid certainly doesn't hurt goals like that—and perhaps explains why Lilly subsequently communicated with greater intelligences and eventually concluded that such intelligence *guided* the universe. It was "as if" Lilly had communicated with extraterrestrials, Verschuur wrote. It was also as if he had communicated with a god. Or both. (Or neither.)

Verschuur didn't want to jump that far into the deep end, but the tanks? Maybe a little weed instead of a lot of LSD? Unfolding every gyrus and crevice in one's own brain? That sounded like a cool afternoon.

"I said to a friend of mine in Boulder, 'If there was an isolation tank, I would definitely go float in one,'" Verschuur told me.

"There is one four blocks away!" said the friend. And of course there was: It was Boulder.

IV

Verschuur knew he couldn't half-ass an experiment like this one. The undertaker had to admit to the possibility that this belief was correct, and just watch what happened next. So he smoked some weed, entered the isolation tank, and became alone. Relieved of the burden of experiencing the world, he settled into the neutral darkness.

So you want to know about other planets, do you? a voice suddenly said.

The sound and content seemed to have come from outside of Verschuur, although that seemed impossible. "Impossible," though, was a descriptor he couldn't use in this kind of experiment. He had to simply experience. And, in fact, that is what the voice said next.

You have to be willing to let go of all your expectations, all your prejudices. The human mind is filled with such barriers, beliefs and expectations about the way the universe is. You want to know about extraterrestrials—then you have to be open to it.

Verschuur, having erected many such barriers during his years in the scientific establishment, asked for assistance with that openness—from who, exactly, he didn't know.

The voice told him to trust. And then it continued on with its strange pseudoscience lesson. Sometimes what humans think of as planets, it revealed, were actually alternate domains of the universe. *Extraterrestrials exist in many more ways than on what you call planets,* it said. *They exist, some of them, in ways you cannot comprehend.*

Soon, the speaker swerved more sharply into metascience, talking inside Verschuur's skull about knowledge creation, experiential methods, and truth—quoting, in fact, John Lilly: *What you believe to be true either is true or becomes true within the limits you can discover through your experience or by experiment. . . . Your experiment will be your experience. Your experience will be your experiment.*

To purposefully make oneself the experiment seems anathema to conventional science. Scientists' *job* is to remove themselves, as much as possible, from their experiments (though they, being human, remain universally incapable). The scientific process is an attempt to strip biases and assumptions away, leaving evidence, correlation, and causation naked.

"Was this what it felt like to go insane?" Verschuur wrote of the experience.

But then, he reminded himself that he had committed to following that experience wherever it would lead. So he went on, dipping himself into the deprivation tank over and over. He even, at one point, bought his own. The voice (whoever it belonged to) became voices. But as the conversations continued, Verschuur grew frustrated with the singularity of his encounters. Whatever was happening, whether he was going crazy or talking to aliens or something else, he could never provide definitive proof to anyone outside this tank. No one could repeat the experience; he couldn't transfer it. He was just like every observer looking out alone at the sky, who can't substantiate that light that appeared and then vanished, leaving no evidence that anything ever happened.

Verschuur didn't like it. If his colleagues heard about these so-called "experiments," they'd surely answer his insanity question in the affirmative. That scared him, for starters. So did the fact that they could be right. So did the idea that the universe and his place in it were, perhaps, very different from what he'd always thought. After all, as soon as he'd dropped the baggage of his beliefs and made room for a new one—that aliens spoke to humans—aliens seemed to speak to him.

"Just because it freaked me out was no reason to leave," he says now.

But being scared to believe is just as real a phenomenon as wanting to believe. In the thirteenth episode of the first *X-Files* season, Mulder and Scully speak to a criminal who claims to have psychic powers. He helps them catch a kidnapper, saves Scully's life, and seems to provide visions to her of her recently deceased father. She explains this all away, in her rational manner. Finally, in the last few minutes of the episode, he offers to deliver one last message to Scully from her father, if she will come to witness his execution.

She does not go. She goes, instead, to Mulder's hospital room, where she rationalizes the seeming powers that the criminal had. Mulder, propped up in the adjustable bed, questions her logic.

"After all you've seen, after all the evidence," he says, "why can't you believe?"

Scully sits down on the bed, her navy blazer a little too close to Mulder's aqua hospital gown. She sighs, stares into the middle distance. Mulder's electrocardiogram beeps in the background, regular as a second hand.

"I'm afraid," she says.

The force of her own words leans her slightly back on the bed. She turns, finally, to look at her partner.

"I'm afraid to believe," she admits.

Partway into his first year of experimentation, Verschuur finally admitted it: "I felt that there were extraterrestrials around," he wrote, "and that the only way they could communicate with humans was through the mind."

He didn't necessarily *believe* that feeling—and he doesn't believe its substance now—but it certainly *felt* true. As a scientist, though, he'd been trained to separate feelings from conclusions: He knew that his emotions did not have the final word on reality.

That's difficult for any human to admit, and especially difficult for some of the people Verschuur met along his journey: a woman at a spiritual outpost in Big Sur who believed she channeled extraterrestrials; a woman convinced an alien had entered her body and subsumed her identity. "Because most scientists are absorbed in their careers, they seldom venture out to explore the magick at least not with all their soul," wrote Verschuur. "Consequently the earnest seekers that wander about in the 'real' world, and I met many individuals in this category, are nonscientists even if they process to be self-critical."

The same is often true in ufology, where witnesses and investigators don't have years of scientific apprenticeship and education, even if they "believe" in the principles of scientific inquiry. That lack leaves them without some of the biases of formal science, of course, but it also leaves them without certain tools that every young scientist is handed. Sometimes their houses of knowledge have the constructional integrity of a self-trained handyman's.

But now to the question we're all here for: Does Verschuur believe in UFOs? Strictly, yes. People see blips in the sky that they can't identify. Interpretively, no. Their inability to explain the blips does not make the blips inexplicable, and definitely doesn't mean they're interplanetary. "I think it's all psychological," he told me. "We can delude ourselves in all sorts of ways."

Before you bristle, "psychological" doesn't carry the same connotations, coming from Verschuur, as it might from another scientist. Verschuur's book spends hundreds of pages detailing his experience seeking out contact in ways most scientists would shake their heads at. His explanations for what he himself experienced fall under "psychological." He's you.

Anyway, he's only seen one UFO himself, on a flight from Denver to Dulles International Airport, while reading a book called *UFOs Explained*. He set his reading material down and pondered why he'd never seen one. And then, just like that, he did: Out the window, a six-inch rod appeared near the engine, traveling the same speed as the jet. A tiny craft, playing with the plane?

Then, though, the jet slowed, turned. The sun struck the area at a different angle. And the rod, once seemingly solid, started to oscillate, then disappeared. It was just an optical illusion. An IFO. What a relief: He would not have to believe.

He stands firm that UFOs, whatever category they are or not eventually identified into, are not alien spacecraft. "There's a lack of imagination of human beings coming up with this hypothesis," Verschuur says.

The universe is stranger than a thing that simply spits out spaceships.

V

What is the universe, though, and how does it actually behave, and why, and who's in it? When will we find out?

We don't really know, and probably we never totally will. We simply take faltering steps along what is probably an asymptote. For Verschuur, explanations about the universe and his experience dip into Jungian psychology.

The brain, Carl Jung said, has buried a *collective* unconscious— snippets of the mind that we all share—deep in the psyche. Our bodies pass it down, earthling to earthling, imprinted with some of the same hopes, beliefs, fears, and behaviors. This shared mind doesn't enter our self-aware awareness, but it appears as common behaviors and archetypes—"psychic structures common to all," as Jung said. Archetypes often manifest themselves as images and motifs. They show up in the stories we tell each other, in the stories we tell ourselves. They become part of religions, of myths.

Jungian psychology is more philosophy than psychology, more pseudoscience than science. But this lens clarified Verschuur's view of the universe. When he went into the isolation tank, he wasn't talking to aliens: The voice was both his own and all of ours. He was talking to himself, the deepest part that he shares with the rest of *this* planet's civilization. Up from those depths floated archetypes, which took the form of words and wisdom from The Others. Alien words from right here on Earth.

Jung himself was no stranger to the extraterrestrial. He actually wrote a book about UFOs, called *Flying Saucers*. The psychologist did not concern himself much with the physicality of the craft. Instead, he focused on what it meant that these saucers showed up, strength in their numbers, right when humans felt most threatened. Regard-less of their material composition or lack thereof, he said, they were undoubtedly a modern myth. An archetype floating up to and down from the sky.

When Verschuur surveys the various attempts at alien contact—from SETI to ufology to close encounters of the high-level kind—he sees the same (archetypal) need he identified early in his alien studies: the urge to believe in something greater than ourselves, greater than this planet. That urge tends to push people toward the same end point. "We're projecting our own image onto the universe," he says.

In the case of UFOs, the nuts-and-bolts technology resembles our experimental military flyers. They are the most competent version of our aerospace industry. Direct alien contact takes the form of speech, concepts, and actions that we can relate to. In religion, gods tend to have hair and muscles. In all three realms, the craft and creatures want to save us or punish us.

When scientists use telescopes to survey the large-scale structure of the universe, they see something interesting: Everywhere you look, it's essentially the same. That's called homogeneity, isotropy. There's not a word for the ufological version of that sentiment, in which UFOs and aliens are—whatever else they are or aren't—mirrors. This holds true whether you believe they come from outer space, whether you believe they're a parapsychological trick, whether you believe they're military, whether you believe they're swamp gas, or whether you believe your fellow humans are crazy.

Even if you only gaze long and hard enough to refute UFOs, what you see is always a version of yourself. You are, in that way, always alone. Even if *we* aren't.

To be clear, Verschuur thinks we *are* alone—practically, if not exactly, speaking. He doesn't *know* that, of course, any more than alien contactees or UFO zealots know ET is on Earth. But that idea may be as salvatory as both the aliens and the gods. If we truly, truly thought that earthlings' problems and fates rested solely in earthlings' hands, would we not live up to the great responsibility that comes with that great power?

"Before we can answer the question 'Is anyone out there?' we must ask and answer another one," Verschuur concluded in his book. "It is this: what is in here?"

VI

Until I was eighteen years old, I was a very religious person, an ultra-devout Mormon. Although I am not now—religious in general or Mormon in particular—I still remember all the things that theology gave me, gift wrapped. The church told me I was chosen and special and that the reason I existed on Earth was so that God could determine my worthiness for heaven. The church also defined "worthiness" for me. All I had to do was what I was told, all I had to believe was what those above me believed. Death was not.

After I had a crisis of faith and left not just this church but all churches, I didn't have any of that knowledge or guidance. I had to grapple with what "be a good person" meant to me—not to some god. I had to peer out at the night sky and, for the first and most devastating time, think that I wasn't here for a reason. "Reason," as a word, didn't even apply. I actually was—we all were—the meaningless speck that my astronomy textbooks had always hinted. Most killer: I would die someday, that would be the end of me, and it wouldn't matter. At all.

I found these personal revelations paralyzing. Early on, driving in a lashing thunderstorm on slick and unseeable roads, I had to pull over because I felt my own mortality for the first time in my whole life. I thought a lot about how nothing in the cosmos gave a shit about me. Nothing I did—nothing scientists or philosophers or philanthropists or dictators did or had ever done—extended in significance beyond the atmosphere of this planet. Or, really, beyond the collective skulls of humanity.

One night, in the midst of this philosophical turmoil and probably procrastinating on my physics homework, I went outside and looked up at the sky. I tried—like really tried—to imagine that all those penlights were actually roiling spheroids made of plasma, scattered across seemingly empty light-years actually filled with invisible matter and energy. I tried to *feel* like this—what I knew to be true—was true. It wasn't easy, in the way that it's not easy to reinhabit any long-lost state of consciousness—new love, adolescent angst, sharp grief—or ones you've just grown so used to that they don't feel like anything anymore—old love, constant angst, dulled grief.

I could only hang on to the actual emotion for a few moments. But during the seconds when I succeeded, I started to feel differently about insignificance. If nothing we do matters to anyone or anything greater than us, if everything—lives, loves, losses, triumphs, terribleness—dies with our civilization, maybe *that* is the part that doesn't matter. All of this human stuff *feels* significant to us, here and now. And maybe that feeling was truer than the truth, that it's probably all for naught.

In writing this book and talking to many members of my own species who I otherwise would not have known, I did not come to believe that flying saucers are real, or that an extraplanetary species made them. I'm not saying there are no aliens that might save or destroy us someday. I'm not even saying they don't drive Earth-bound spaceships. I don't doubt that people have seen things in the sky they can't explain. I admit the possibility, remote though it may be, that *no* human could explain those things.

So while no evidence or experience convinced or converted me, I did come to see the cosmos differently. I view it the way I did when I was a kid: as if I were discovering it for the first time, as if its high-up strangeness were a revelation, and as if I would never fully get it, no matter how hard I tried.

On Earth, scientists declare paper by paper their comprehension of a new phenomenon, their discovery of a new object, their synthesis of the laws of the universe. But their comprehension is piecemeal—coming in snippets that have never cohered into a whole and probably never will.

Do we as a civilization know much more than we used to? Sure.

Are we also very, very wrong about things we currently feel certain of? Surely. Just as surely as we were wrong about how gods exploded volcanoes, and how making someone bleed healed them.

The grand scope of the universe may always remain beyond our grasp. Take this: What existed before the Big Bang? What existed before the before of the Big Bang? Where did the very first matter and energy come from? What can "very first" even mean in this context if linear time didn't even exist?

Such topics may stay forever beyond the bounds of empirical science. We can only observe what we can observe. We can only comprehend things our brains are wired to conceive of. Fundamental limits may exist to what we can even try to know. And whatever those fundamental limits are, humans aren't even close to hitting them. But because we're humans, we'll keep trying, at least until we blow each other up or heat ourselves to death. In that quest, we'll undoubtedly have to ditch lots of barriers, beliefs, and expectations that we currently count as knowledge.

At one point in his book, Verschuur borrows some language from *Hamlet*, a play in which the main character feigns madness and—depending on whose high school essay you read—might actually lose his mind at some point in the plot. "I believe," Verschuur wrote, "that our 'hope' lies not in searching for and finding civilizations such as our own but in becoming open to the possibility that there is more in our universe than we have yet dreamt in our philosophies."

That, I think, is the one thing I know, feel, and believe in: universal uncertainty. Humans are so far from understanding the what, where, when, why, how, and who of our swath of spacetime. And the truth of the future is likely much stranger than the fictions of the present.

ACKNOWLEDGMENTS

I did not write this book by myself. For their contributions, I'd like to thank, first, everyone who shared their stories and insights with me during the reporting process, or unwittingly allowed me to peer over their keyboards at what they typed onto forums. Even if your words didn't make it into the final text, they've stuck in my mind and helped shape what appears between these covers. No one has to agree to the personal invasion that is the interview process, but I appreciate very much that you did.

For going on a googolplex of ufological road trips, listening to the play-by-play of my revelations or lack thereof, and for putting up with the song "Aliens Exist" on repeat, I'd like to thank my sister Rebekah Scoles. To my parents, Ron and Darla, and my other sister, Rachel, thank you for the consistent encouragement to do weird things and the framework for thinking about belief systems. To the Scoles and Kinney extended families, thanks for enduring strange holiday-time conversation topics.

Brooke Napier, I'm too generally grateful for you to have much to say specifically about it, but thanks for being born and for trying out the CE-5 protocols that one time. Ann Martin, I hope you have a clone on another planet, because the universe deserves two of you. Carolyn Belle, Tasha Eichenseher, Patrick Kadel, Hannah Scarborough, and Tripp Jones, thank you for your mostly unconditional friendship, your gameness, your openness, and your willingness to speculate over coffee, beer, mountaintops, and hours.

The teachers of the Governor's STEM Institute in West Virginia provided an invaluable, if profane, sounding board and support system. Ladies Who Launch: You know who you are. And, Katie Palmer, thanks for being the editor who said, of the initial AATIP coverage, "Want to dig in?"; for understanding what I mean to say better than I do; and for being the Scully to all my out-there ideas.

Thanks go also, of course, to Jessica Case and Pegasus Books for publishing this one, and to my agent, Zoe Sandler, who made this book and everything else better than it would have been otherwise.

SELECTED REFERENCES

4:20 TV FREEDOMIST FILMS. "Sunspot What's Happening?" *YouTube*, YouTube, 13 Sept. 2018, www.youtube.com/watch?v=_AVdVBvSjcU.*

"Advanced Aerospace Weapon System Applications Program—Solicitation HHM402-08-R-0211." *Fed Biz Opps*, 2008, http://www.fbodaily.com/archive/2008/08-August/20-Aug-2008/FBO-01643684.htm.

"Aerial Whatzits Buzz D.C. Again." *Washington Daily News*, 1952.

Alberini, Cristina M., et al. "Memory Reconsolidation." *Current Biology*, 2013, pp. 81–117.

Associated Press. "Millionaire Searches for UFOs on Ranch in Utah." *The Register-Guard* 24 Oct. 1996, https://news.google.com/newspapers?nid=1310&dat=19961024&id=O0xWAAAAIBAJ&sjid=HOsDAAAAIBAJ&pg=4548,6235898.

Arnold, Kenneth, and Raymond Palmer. *The Coming of the Saucers.* CreateSpace Independent Publishing Platform, 2014.

Arnu, Joerg. "Dreamland Resort." http://www.drealandresort.com.

Assange, Julian. "Emails Including the Term 'Tom DeLonge.'" *WikiLeaks*, https://search.wikileaks.org/?query=tom+delonge&exact_phrase=&any_of=&exclude_words=&document_date_start=&document_date_end=&released_date_start=&released_date_end=&new_search=True&order_by=most_relevant#results.

Bader, Christopher, et al. *Paranormal America: Ghost Encounters, UFO Sightings, Bigfoot Hunts, and Other Curiosities in Religion and Culture.* NYU Press, 2011, https://www.amazon.com/Paranormal-America -Encounters-Sightings-Curiosities/dp/0814791352/ref=sr_1_1?keywor ds=paranormal+america&qid=1563943836&s=gateway&sr=8-1.

Basterfield, Keith. "The BAASS Team—Some of Their Roles and Some of Their Names." *Unidentified Aerial Phenomena—Scientific Research,* http://ufos-scientificresearch.blogspot.com/2018/05/the-baass-team -some-of-their-roles-and.html.

Battaglia, Debbora, ed. *E.T. Culture: Anthropology in Outerspaces.* Duke University Press Books, 2006.

Bender, Bryan. "The Pentagon's Secret Search for UFOs." *POLITICO Magazine,* http://politi.co/2CH9Q2f.

———. "U.S. Navy Drafting New Guidelines for Reporting UFOs." *POLITICO Magazine,* https://politi.co/2USYNjd.

Brewer, Jack. "The Carpenter Affair: For the Record." *The UFO Trail,* 22 Oct. 2017, http://ufotrail.blogspot.com/2013/10/the-carpenter-affair -for-record.html.

———. *The Greys Have Been Framed: Exploitation in the UFO Community.* CreateSpace Independent Publishing Platform, 2015.

———. "AATIP Crew Handled Kimbler Roswell Debris." *The UFO Trail,* 28 June 2018, http://ufotrail.blogspot.com/2018/06/aatip-crew -handled-kimbler-roswell.html.

———. "What Happened to the Ambient Monitoring Project?" *The UFO Trail,* 2 Apr. 2014, http://ufotrail.blogspot.com/2014/04/what -happened-to-ambient-monitoring.html.

Bryan, C. D. B. *Close Encounters Of The Fourth Kind: Alien Abduction, UFOs, and the Conference at M.I.T.* 1st edition, Knopf, 1995.

Butrica, Andrew. *To See the Unseen: A History of Planetary Radar Astronomy.* NASA History Office, 1996.

Campbell, Glenn. "Area 51 Viewers' Guide." 1995.

———. *The Groom Lake Desert Rat.*

———. "UFOMIND: Bigelow Group Reveals Personnel." 11 Oct. 1998, http://www.ufoupdateslist.com/1998/oct/m12-009.shtml.

Carrion, James. "Strange Bedfellows." *The UFO Chronicles*, 3 Feb. 2011, https://www.theufochronicles.com/2011/02/ufo-news-former-mufon-director-james.html.

———. *The Roswell Deception*. 2018.

Clarkson, James. "With Regret—Why I Must Leave MUFON Completely." *James Clarkson: UFO Investigations*, 22 July 2017, http://web.archive.org/web/20170729173459/http://jamesclarksonufo.com/ufo-news/with-regret-why-i-must-leave-mufon-completely.

Colavito, Jason. "Top MUFON Official Quits Over Organization's Continued Support of John Ventre a Year After Ventre's Racist Rant." *Jason Colavito*, 17 Apr. 2018, http://www.jasoncolavito.com/blog/top-mufon-official-over-organizations-continued-support-of-john-ventre-a-year-after-ventres-racist-rant.

Condon, Edward. *Scientific Study of Unidentified Flying Objects; Final Report of Research Conducted by the University of Colorado for the Air Force Office of Scientific Research under the Direction of Edward U. Condon.* Bantam, 1969.

Cooper, Helene, et al. "Glowing Auras and 'Black Money': The Pentagon's Mysterious U.F.O. Program." *The New York Times*, 16 Dec. 2017, https://www.nytimes.com/2017/12/16/us/politics/pentagon-program-ufo-harry-reid.html.

Copeland, Preston. "Saucers and the Sacred: The Folklore of UFO Narratives." *Utah State University*, May 2012, https://digitalcommons.usu.edu/gradreports/149.

Corbell, Jeremy. *Bob Lazar: Area 51 & Flying Saucers.* The Orchard, 2018, https://www.amazon.com/Bob-Lazar-Area-Flying-Saucers/dp/B07L1DD8L6/ref=sr_1_1?keywords=bob+lazar+documentary&qid=1563943445&s=gateway&sr=8-1.

———. *Watch Hunt for the Skinwalker.* The Orchard, 2018, https://www.amazon.com/Hunt-Skinwalker-George-Knapp/dp/B07H5SS1F2/ref=sr_1_2?keywords=%22Jeremy+Kenyon+Lockyer+Corbell%22&qid=1563943532&s=instant-video&sr=1-2.

Darrach, H. B., and Robert Ginna. "Have We Visitors from Outer Space." *LIFE.*

DD Form 1910 for Videos "GoFast," "Gimble," and "FLIR." Defense Office of Prepublication and Security Review, 24 Aug. 2017, https://www .theblackvault.com/casefiles/wp-content/uploads/2019/02/redacted -Clearance-Request_1556576605509_85063552_ver1.0-1.jpg.

Dean, Jodi. *Aliens in America: Conspiracy Cultures from Outerspace to Cyber-space.* 1 edition, Cornell University Press, 1998.

Defense Intelligence Agency. *Department of Defense, Message Center, Iran Incident.* 1976.

Delonge, Tom, and A. J. Hartley. *Sekret Machines Book 1: Chasing Shadows.* To the Stars, 2017, https://www.amazon.com/Sekret-Machines-Book -Chasing-Shadows/dp/1943272298/ref=sr_1_1?keywords=sekret+mac hines&qid=1563943941&s=gateway&sr=8-1.

DeLonge, Tom, and A. J. Hartley. *Sekret Machines Book 2: A Fire Within.* To the Stars, 2018, https://www.amazon.com/Sekret-Machines-Book -Fire-Within/dp/1943272344/ref=sr_1_2?keywords=sekret+machines &qid=1563943941&s=gateway&sr=8-2.

Dent, Steven. *CIA Dragonfly Drone Almost Beat Modern UAVs by 40 Years.* 30 July 2012, https://www.engadget.com/2012/07/30 /cia-dragonfly-drone-uavs-40-years/.

Department of the Air Force. *Air Force Regulation No. 200-2.* 5 Feb. 1958, http://www.nicap.org/directives/AFR%20200-2,%20Feb%20 5,%201958.pdf.

Dodd, Adam. "Strategic Ignorance and the Search for Extraterrestrial Intelligence: Critiquing the Discursive Segregation of UFOs from Scientific Inquiry." *Astropolitics*, vol. 16, no. 1, Apr. 2018, pp. 75–95.

Dorsch, Kate. *Seeing Is Believing: A Historical Perspective on the Ontological Status of UFOs (Abstract)—July 2016.* www.academia.edu, https: //www.academia.edu/27070724/Seeing_is_Believing_A _Historical_Perspective_on_the_Ontological_Status_of_UFOs _Abstract_-_July_2016.

Edwards, Gavin. "Blink-182: The Half-Naked Truth." *Rolling Stone*, 3 Aug. 2000, https://www.rollingstone.com/music/music-news/blink -182-the-half-naked-truth-87106/.

———. "Punk Guitar + Fart Jokes = Blink-182." *Rolling Stone*, Jan. 2000, https://www.rollingstone.com/music/music-news/punk-guitar-fart-jokes-blink-182-63042/.

FAA. *Air Traffic Organization Policy: Unidentified Flying Object (UFO) Reports*. 2011, http://tfmlearning.faa.gov/publications/atpubs/ATC/atc0908.html.

Fuentes, Gerald. "A Report on the Roper Analysis Data." *ViewZone*.

Graham, Robbie. *UFOs: Reframing the Debate*. White Crow Books, 2017.

Greenewald, John. "To The Stars Academy of Arts & Science Archives." *The Black Vault Case Files*, https://www.theblackvault.com/casefiles/tag/to-the-stars-academy-of-arts-science/.

Greenewald Jr., John. *Beyond UFO Secrecy*. 2nd edition, Galde Press, 2008.

Greer, Steven. "The CE-5 Initiative." https://siriusdisclosure.com/wp-content/uploads/2012/12/CE-5-Initiative-Transcript.pdf.

Grush, Loren. "Space Industry CEO Is 'Absolutely Convinced' that Aliens Have Visited Earth." *The Verge*, 30 May 2017, https://www.theverge.com/2017/5/30/15712270/robert-bigelow-ufo-aliens-60-minutes-aerospace.

Haines, Gerald. "CIA's Role in the Study of UFOs, 1947–90." *Studies in Intelligence*, vol. 1, no. 1, 1997.

"Harassed Rancher Who Located 'Saucer' Sorry He Told about It." *Roswell Daily Record*, 9 July 1947.

Hoyt, Diana Palmer. *Ufocritique: Ufos, Social Intelligence, and the Condon Report*. Virginia Tech, 20 Apr. 2000. *vtechworks.lib.vt.edu*, https://vtechworks.lib.vt.edu/handle/10919/32352.

Hughes, Josiah. "Tom DeLonge Says He's Doing 'Really, Really Important Things' with Area 51." *Exclaim*, http://exclaim.ca/music/article/tom_delonge_says_hes_doing_really_really_important_things_with_area_51.

Hynek, J. Allen. *The UFO Experience*. 5th printing edition, Ballantine Books, 1977.

"Impossible, Maybe, But Seein' Is Believin'." *East Oregonian*, 25 June 1947.

"In Focus: Eyewitness Misidentification." Innocence Project, 21 Oct. 2008, www.innocenceproject.org/in-focus-eyewitness-misidentification/.s

Jacobs, David M. *The UFO Controversy in America*. 1st Edition, Indiana University Press, 1975.

Jacobsen, Annie. *Area 51: An Uncensored History of America's Top Secret Military Base*. Large print edition, Little, Brown and Company, 2011.

"Jets Chase D.C. Sky Ghosts." *New York Daily News*, 1952.

Johnson, Dirk. "Coloradans Raise Voices against Jet Fighters' Din." *New York Times*, 22 June 1992, https://www.nytimes.com/1992/06/22/us/coloradans-raise-voices-against-jet-fighters-din.html?mtrref=undefined&gwh=6867DB428F6EC4EDA0780A9613C57EE4&gwt=pay.

Jung, Carl. *Personal Correspondence to Gilbert C. Harrison*. 12 Dec. 1957.

Kapnisi. *Letter to Committee on Armed Services, Listing All Products Produced under AATIP*. 9 Jan. 2018.

Kelleher, Colm, and George Knapp. *Hunt for the Skinwalker: Science Confronts the Unexplained at a Remote Ranch in Utah*. Paraview Pocket Books, 2005, https://www.amazon.com/Hunt-Skinwalker-Science-Confronts-Unexplained/dp/1416505210/ref=sr_1_2?keywords=hunt+for+the+skinwalker&qid=1563943548&s=gateway&sr=8-2.

Key, William. "'Sky Devil-Ship' Scares Pilots; Air Chief Wishes He Had One." *Atlanta Journal-Constitution*, 25 July 1948.

Kimble, Valerie. "Socorro's UFO Incident Still Unexplained." *El Defensor Chieftain*, 23 July 2003.

Verschuur PhD, Gerrit L. *Is Anyone Out There?: Personal Adventures in Search for Extraterrestrial Intelligence*. CreateSpace Independent Publishing Platform, 2015.

Lang, Richard. "What Caused the Failure of the BAASS–MUFON SIP Program?" *The UFO Chronicles*, 6 Mar. 2011, https://www.theufochronicles.com/2011/03/what-caused-failure-of-baass-mufon-sip.html.

Lepselter, Susan. *The Resonance of Unseen Things: Poetics, Power, Captivity, and UFOs in the American Uncanny*. University of Michigan Press, 2016.

Lichtensteev, Grace. "11 States Baffled by Mutilation of Cattle." *New York Times*, 30 Oct. 1975, https://www.nytimes.com/1975/10/30/archives/11-states-baffled-by-mutilation-of-cattle.html.

Loftus, Elizabeth F, and Jacqueline E Pickrell. "The Formation of False Memories." *Psychiatric Annals*, vol. 25, no. 12, 1995, pp. 720–725.

Low, Robert. *Some Thoughts on the UFO Project.* 9 Aug. 1966, http://www.nicap.org/docs/660809lowmemo.htm.

Melley, Timothy. *Empire of Conspiracy: The Culture of Paranoia in Postwar America.* 1st edition, Cornell University Press, 2000.

Mellon, Christopher. "The Military Keeps Encountering UFOs. Why Doesn't the Pentagon Care?" *Washington Post*, 9 Mar. 2018. *www.washingtonpost.com*, https://www.washingtonpost.com/outlook/the-military-keeps-encountering-ufos-why-doesnt-the-pentagon-care/2018/03/09/242c125c-22ee-11e8-94da-ebf9d112159c_story.html.

Messoline, Judy. *That Crazy Lady Down the Road.* 1st edition, Earth Star Publications, 2005.

Monroe, Jazz. "Tom DeLonge on Blink-182, UFOs and Meeting with US Air Force Space Command." *NME*, 2 Sept. 2015, https://www.nme.com/blogs/nme-blogs/tom-delonge-on-blink-182-ufos-and-meeting-with-us-air-force-space-command-764561.

Neisser, Ulric, and Nicole Harsch. "Phantom Flashbulbs: False Recollections of Hearing the News about Challenger." *Affect and Accuracy in Recall*, 1992, pp. 9–31.

O'Brien, Christopher. *UFO Data Acquisition Project.* https://ufodap.com/p/slv/overview.

O'Connell, Mark. *The Close Encounters Man: How One Man Made the World Believe in UFOs.* Dey Street Books, 2017.

Pope, Nick. *Open Skies, Closed Minds.* Thistle Publishing, 2015.

Poulsen, Kevin. *Area 51 Hackers Dig up Trouble.* 25 May 2004.

Printy, Tim. "Brazel Debris Field Imagery." *SUNlite*, vol. 4, no. 6, 2012.

Project BLUE BOOK—Unidentified Flying Objects. National Archives, https://www.archives.gov/research/military/air-force/ufos.html.

"Project Grudge Documents." *Air Force Declassification Office*, http://www.secretsdeclassified.af.mil/News/Article-Display/Article/459840/project-grudge-documents.

"RAAF Captures Flying Saucer on Ranch in Roswell Region," *Roswell Daily Record,* 8 July 1947.

Randle, Kevin, and Donald Schmitt. *UFO Crash at Roswell*. Avon, 1991, https://www.amazon.com/Ufo-Crash-Roswell-Kevin-Randle /dp/0380761963/ref=sr_1_1?keywords=ufo+crash+at+roswell&qid=156 3944194&s=gateway&sr=8-1.

"Real Estate Mogul Reaches For the Stars." *Wall Street Journal*, 25 Aug. 1999.

Redfern, Nick. "Alien Contact in Puerto Rico?" *Mysterious Universe*, https://mysteriousuniverse.org/2014/08/alien-contact-in-puerto-rico/.

Riley, Albert. "Atlanta Pilots Report Wingless Sky Monster." *Atlanta Journal-Constitution*, 25 July 1948.

Robertson Panel. *Report of Scientific Advisory Panel on Unidentified Flying Objects Convened by Office of Scientific Intelligence, CIA*. Central Intelligence Agency.

Rojas, Alejandro. "Roswell Alien Slides Photo Revealed." *Open Minds*, 6 May 2015, http://www.openminds.tv/roswell-alien-slides-photo-re vealed/33494.

Ruppelt, Edward J. *The Report On Unidentified Flying Objects*. CreateSpace Independent Publishing Platform, 2011.

Saad, Lydia. "Americans Skeptical of UFOs, but Say Government Knows More." *Gallup.com*, Gallup, 6 Sept. 2019, news.gallup.com /poll/266441/americans-skeptical-ufos-say-government-knows .aspx.

"Sacramento Peak Draft Environmental Impact Statement." *National Science Foundation*, 2018, https://www.nsf.gov/mps/ast/env_impact _reviews/sacpeak/sacpeak_drafteis.jsp.

Saler, Benson, et al. *UFO Crash at Roswell: The Genesis of a Modern Myth: Benson Saler, Charles A. Ziegler, Charles Moore*. Smithsonian Books, 2010, https://www.amazon.com/UFO-Crash-Roswell-Genesis-Modern /dp/1588340635/ref=sr_1_2?keywords=ufo+crash+at+roswell&qid=15639 44136&s=gateway&sr=8-2.

"'Saucer' Outran Jet, Pilot Reveals." *Washington Post*, 28 July 1956.

SEC Filings for To The Stars Academy. Securities and Exchange Commission, https://www.sec.gov/cgi-bin/browse-edgar?company=to+the+star s+academy&match=contains&action=getcompany.

Seedhouse, Erik. *Bigelow Aerospace: Colonizing Space One Module at a Time*. Springer International Publishing, 2015. *www.springer.com*, https://www.springer.com/gp/book/9783319051963.

Shalett, Sidney. "What You Can Believe about Flying Saucers." *The Saturday Evening Post*, 7 May 1949.

Sheaffer, Robert. "Bad UFOs: Skepticism, UFOs, and The Universe: 'Roswell Crash Debris'—Claims and More Claims." *Bad UFOs*, 2 July 2018, https://badufos.blogspot.com/2018/07/roswell-crash-debris-claims-and-more.html.

——. "Bigelow's Aerospace and Saucer Emporium." *Skeptical Enquirer*, July 2009, https://skepticalinquirer.org/2009/07/bigelows_aerospace_and_saucer_emporium/.

Shields, Henry. *Now You See It, Now You Don't*. HQ USAFE/INOMP.

Shostak, Seth. "Why Solar Observatory's Mysterious Closure Sparked Talk of Aliens." *SETI Institute*, 21 Sept. 2018, https://www.seti.org/why-solar-observatorys-mysterious-closure-sparked-talk-aliens.

Shough, Martin. "The Singular Adventure of Mr. Kenneth Arnold," National Aviation Reporting Centre on Anomalous Phenomena, 2010.

"Supersonic Flying Saucers Sighted by Idaho Pilot." *Chicago Sun*, 26 June 1947.

Tedder, Michael. "Blink-182 Co-Founder Tom DeLonge Goes Deep on UFOs, Government Coverups and Why Aliens Are Bigger than Jesus." *PAPER*, Feb. 2015, http://www.papermag.com/blink-182-co-founder-tom-delonge-goes-deep-on-ufos-government-coverups-1427513207.html.

Twining, Nathan. "AMC Opinion Concerning 'Flying Discs.'" 23 Sept. 1947.

——. "Letter from General N. F. Twining to Commanding General, Army Air Force." 23 Sept. 1947.

Unidentified Aerial Objects: Project "SIGN." Headquarters, Air Materiel Command, Wright-Patterson Air Force Base.

US Air Force. *Report of Air Force Research Regarding the "Roswell Incident."* 1994.

——. *Roswell Report: Case Closed*. Military Bookshop, 2011.

Van Eyck, Zack. "Frequent Fliers." *Deseret News*, 30 June 1996, https://www.deseretnews.com/article/498676/FREQUENT-FLIERS.html.

——. "Millionaire Leads Quest for UFO Data." *Deseret News*, 20 Oct. 1996, https://www.deseretnews.com/article/520340/MILLIONAIRE-LEADS-QUEST-FOR-UFO-DATA.html.

——. "No UFOs or ETs Have Dropped in at Spooky Ranch." *Deseret News*, 27 Apr. 1997, https://www.deseretnews.com/article/557318/No-UFOs-or-ETs-have-dropped-in-at-spooky-ranch.html.

——. "Private UFO Study Takes a Public Turn." *Deseret News*, 10 Aug. 1998, https://www.deseretnews.com/article/646095/Private-UFO-study-takes-a-public-turn.html.

"Watershed: The Chiles-Whitted 'Rocketship' Sighting." *Project 1947*, http://www.project1947.com/gr/chileswhitted.htm.

Whalen, Andrew. "What If Aliens Met Racists? MUFON Resignations Highlight Internal Divisions in UFO Sightings Organization." *Newsweek*, 29 Apr. 2018, https://www.newsweek.com/ufo-sightings-mufon-2018-john-ventre-alien-extraterrestrial-905060.

Zapped by Hostile Space Aliens. 1988, https://public.nrao.edu/wp-content/uploads/2013/09/gallery-images-large300ftzappedpaper_large.jpg.

INDEX